MW00883195

Summary

From Algebra to Graphs

Transforming Relational Data for the Future

An exciting book that reveals the secrets of graph algebra and its practical applications in today's world. Learn to master the key concepts of graph algebra and graph theory, explore innovative techniques for analyzing and exploiting relational data through concrete and captivating examples.

Whether you are a seasoned mathematician, a data enthusiast, or simply curious about a new approach to solving complex problems, you will dive into areas such as social network analysis, recommender systems, and bioinformatics to understand how graph algebra shapes the world around us.

Don't miss this opportunity to expand your horizons and discover the wonders of graph algebra!

Aymeric NEUVY,

Professor of Mathematics, Data Engineer graduated in 2023 from Ecole Centrale d'Electronique de Paris, with additional training at University of California, Los Angeles and OMNES, San Francisco - USA. Speaker in Machine Learning, Blockchain, works as a Data Engineer in an IT consulting company

Chapter 1: Introduction

1) Purpose of the book

The world we live in is complex and fascinating, governed by laws and principles that can be explored and understood through mathematics. This book is an invitation to discover the algebraic structures behind graphs and relational data. The main goal of this book is to present the key concepts of graph algebra and their application to relational data processing, with an emphasis on intuitive understanding of the underlying ideas rather than on mastering the technical details. We aim to show how algebraic structures and graph theoretical techniques can be used to solve concrete problems in various domains, such as social network analysis, recommender systems and bioinformatics (such as DNA sequence interpretation, genomic data, protein structure,...).

By approaching the topics in a progressive and pedagogical way, we aim to make mathematics accessible and attractive to a wide audience, ranging from students to researchers and professionals working in the field of relational data processing. Our goal is to share the passion and enthusiasm for mathematics, highlighting the amazing connections between algebra, graphs and practical real-world applications. To achieve this goal, we have structured this book into several chapters that cover fundamental aspects of algebra and graphs, graph algebra concepts, relational data processing algorithms and techniques, and practical applications and future challenges in this field. Each chapter is designed to be read independently, but we encourage readers to follow the suggested order for a better overall understanding of the concepts and techniques presented. Whether you are a student, researcher, or professional looking to improve your relational data processing skills, we hope that this book will inspire you and help you develop your mathematical intuition and ability to solve real-world problems using the tools and techniques of graph algebra.

2) Chapter Overview

In this book, we will explore in depth the fascinating aspects of graph algebra and relational data processing. Chapter 2 will introduce you to the fundamentals of algebra and graphs. We will introduce the key concepts of algebra, such as groups, rings and bodies, as well as linear algebra. Then we will turn to graph theory, exploring the basic notions of graphs, nodes and edges, as well as the different types of graphs and their properties. We will end this chapter by examining representations of graphs.

In Chapter 3, we will dive into the world of graph algebra. We will study matrices associated with graphs, such as adjacency, incidence and Laplacian matrices. We will also discuss algebraic operations on graphs, such as the sum and product of graphs, and powers of graphs. Finally, we will discuss characteristic polynomials and the spectrum of graphs.

Chapter 4 will focus on algorithms and techniques for processing relational data. We will explore graph mining methods, including depth-first and breadth-first traversals, and Dijkstra and Bellman-Ford algorithms. We will also discuss graph partitioning techniques, introducing spectral and flow methods. This chapter will conclude with an overview of graph optimization problems and their associated algorithms, such as coupling algorithms, the shortest path problem and the traveling salesman problem.

Chapter 5 will discuss practical applications of graph algebra in various fields. We will start by looking at social network analysis and centrality measures, as well as community detection. Next, we will look at recommender systems, presenting collaborative filtering and content-based models. Finally, we will explore applications of graph algebra in bioinformatics and genomics, including sequence alignment.

Chapter 6 will discuss challenges and future prospects in graph algebra and relational data processing. We will discuss scalability and processing of massive graphs, machine learning on graphs, and visualization and interpretation of complex graphs. Finally, Chapter 7 will conclude this book by summarizing key concepts and highlighting the implications and impact of graph algebra on relational data processing. We will summarize the main ideas presented in each chapter and highlight the importance of graph algebra in understanding and solving complex problems in the world today. This book is the result of a passionate effort to share knowledge and enthusiasm for mathematics, graph algebra, and relational data processing. We hope that this in-depth exploration of concepts, techniques, and practical applications will inspire you and help you develop your mathematical intuition and ability to solve real-world problems in your field of study or work.

Chapter 2: Fundamentals of algebra and graphs

1) Key concepts of algebra

In this first section of Chapter 2, we will explore the foundations of algebra and introduce the key concepts that will serve as the basis for our later study of graph algebra. We will cover the notions of groups, rings, bodies, and linear algebra, emphasizing intuition and understanding of ideas rather than mastery of technical details. We will begin our exploration of algebra with the notion of a group, which is a fundamental algebraic structure consisting of a set of elements and an operation that combines those elements. A group is characterized by four properties: closure, associativity, existence of a neutral element and existence of inverses.

*Closure: for any elements a and b in the group G, the binary operation * (multiplication) is also defined and produces a result that is also in G. In other words, if a and b are elements of the group G, then a*b is also an element of G.*

*Associativity: for all elements a, b and c in the group G, the binary operation * is associative, i.e. (ab)c = a(bc).*

The existence of a neutral element: there exists an element e in the group G such that for any element a of G, ae = ea = a. This element is called the neutral element or the identity.

The existence of inverses: for any element a in G, there exists an element b in G such that ab = ba = e (the neutral element). The element b is called the inverse of a and is noted a^{-1}

We will illustrate these properties with concrete examples, such as groups of integers modulo n and groups of permutations, and we will discuss the importance of groups in the study of symmetries and transformations. Next, we will discuss the notion of a ring, which is an extension of the group concept and includes a second operation, called multiplication, in addition to the addition operation. A ring is defined by a set of elements and two operations that satisfy certain properties, such as the distributivity of multiplication over addition. We will present examples of rings, such as rings of integers and rings of polynomials, and emphasize their central role in the study of algebraic structures and equations. The notion of a body, which is an even richer algebraic structure than rings, will be our third step in this exploration of the key concepts of algebra. A body is a ring in which all non-zero elements have a multiplicative inverse, which means that we can perform divisions without restriction. Examples of bodies include the rationals, reals and complexes, as well as finite bodies, which play a crucial role in cryptography and code theory. We will discuss the importance of bodies in the study of equations and vector spaces, and show how they generalize notions of calculus and geometry. Finally, we will look at linear algebra, which is the study of vector spaces and linear transformations between them. Vector spaces are sets of elements called vectors, which can be added and multiplied by scalars, i.e. elements of a body. Linear transformations are functions that preserve the operations of addition and multiplication by a scalar. We will introduce the basic concepts of linear algebra, such as vectors, matrices, vector spaces, subspaces, linear dependence and independence, and bases and dimensions. We will also explore matrix operations, such as addition, multiplication, and inversion of matrices, as well as systems of linear equations and solution methods, including Gauss elimination and LU decomposition.

In this chapter, we will highlight the connections between algebraic concepts and real-world problems, and show how algebra provides a unifying language for describing and solving a wide variety of problems in fields such as physics, computer science, economics, and biology. In the remainder of this chapter, we will discuss graph theory, which is a branch of mathematics closely related to algebra and plays a central role in the study of networks and relationships. We will explore the basic notions of graphs, nodes and edges, as well as the different types of graphs and their properties, and we will examine graph representations and manipulation algorithms. This introduction to graph theory will provide you with the tools and concepts necessary to understand and apply graph algebra in relational data processing and various application areas.

This subsection is devoted to the study of rings, algebraic structures that combine aspects of addition and multiplication. Rings are characterized by several properties, which give them a deep and nuanced structure. First, rings must satisfy the four properties of groups, with the addition of the commutativity of addition. Second, multiplication must be associative and distributive with respect to addition. Unlike groups, rings do not require that multiplication be commutative, nor that there be a neutral element for multiplication or multiplicative inverses. The study of rings provides a unifying language for understanding the interactions between addition and multiplication, and their applications extend far beyond the boundaries of mathematics, touching areas such as physics, computer science, and cryptography. In this survey, we will explore the fundamental properties of rings, presenting their characteristics and their differences from groups. We will also study ideals, which are subsets of a ring with specific properties, as well as the notions of quotient rings and ring homomorphisms. These notions will help us better understand the structure and diversity of rings. Rings can take many forms and appear in many contexts. Rings of integers and rings of polynomials are two examples of rings commonly used in mathematics, but there are many others.

Linear algebra is a fascinating and compelling area of mathematics that studies vector spaces and linear transformations. It occupies a central place in many areas of mathematics, physics, and engineering, and is particularly relevant to our exploration of the connections between algebra and data engineering. Vector spaces are sets of objects, called vectors, that are subject to addition and multiplication operations by scalars. They are defined by a number of properties, such as closure, associativity, distributivity, and the existence of a neutral element and inverses. Vector spaces can be finite or infinite dimensional, and can take many shapes and sizes. Classic examples of vector spaces include Euclidean spaces, such as , and function spaces. Bases and dimensions are essential concepts in linear algebra. A basis of a vector space is a set of linearly independent vectors that generate the entire space. The dimension of a vector space is the number of vectors in its basis, and it is an important invariant that characterizes the size and complexity of the space. The coordinates of a vector with respect to a given basis are the coefficients that allow it to be written as a linear combination of the vectors in the basis.

An example of a vector space is the vector space of polynomials of degree less than or equal to 2, denoted $\mathbb{P}_2$.

It consists of all polynomials of the form $ax^2 + bx + c$*, where a, b and c are real scalars. The vectors in this space are therefore polynomials. The operations of addition and multiplication by a scalar are defined in the usual way for polynomials. The canonical basis of* $\mathbb{P}_2$ *is the set* $1, x, x^2$ *. This basis consists of three vectors which are linearly independent and which generate the whole vector space. In other words, any polynomial of degree less than or equal to 2 can be written as a linear combination of these three vectors. The coordinates of a polynomial vector* $p(x) = ax^2 + bx + c$ *with respect to this base are the coefficients a, b and c. Thus, we can write p(x) in the following form:*

$$p(x) = ax^2 + bx + c = \begin{pmatrix} a & b & c \end{pmatrix} \cdot \begin{pmatrix} x^2 \\ x \\ 1 \end{pmatrix}$$

Linear transformations are applications between vector spaces that preserve the operations of addition and multiplication by scalars. They are the central objects of study in linear algebra, and they allow us to model a wide variety of phenomena and processes, such as rotations, expansions, projections, and systems of linear equations. Linear transformations can be represented by matrices, which are rectangular arrays of numbers organized into rows and columns. Matrices allow us to manipulate and solve linear transformations in an efficient and systematic way. The concepts of rank, kernel and image are also essential in linear algebra. The rank of a matrix is the maximum number of linearly independent columns (or rows), and it is related to the dimension of the image of a linear transformation. The kernel of a linear transformation is the set of vectors that are sent to the neutral element, and it is a vector subspace of the starting space. The image of a linear transformation is the set of vectors that are reached by the application, and it is a subspace of the target space.

2) Graph theory

Graph theory is a branch of mathematics that studies objects called graphs, which are made up of vertices (or nodes) and edges (or links) that connect them. It was born from the curiosity and imagination of Euler, who sought to solve the famous Königsberg bridge problem (*). Since then, the theory of graphs has experienced a remarkable growth and has become a central field of mathematics, with applications in various disciplines such as computer science, biology, sociology, and of course, data engineering. A graph is a set of vertices and edges, which can be represented schematically by points and lines connecting them. Edges can be directed or undirected, depending on whether or not they have an orientation. Graphs can be finite or infinite, simple or complex, and they can have various properties and structures, such as cycles, cliques, related components and induced subgraphs. Graphs are extremely versatile and expressive tools for representing and analyzing a wide variety of situations and problems. Graph theory is rich in concepts and methods, which are intended to study and characterize graphs in a rigorous and systematic way. Among these concepts, we find the notion of degree of a node, which is the number of edges that are incident to it, and the notions of path, circuit and distance, which allow us to describe the relations and interactions between the nodes of a graph. The global properties of graphs, such as connectedness, planarity, coloring and isomorphism, are also central objects of study in graph theory. Graph theory also encompasses a wide range of techniques and algorithms, which are designed to solve problems and optimize criteria on graphs. Among these techniques are traversal algorithms, such as breadth-first and depth-first traversal, which allow graphs to be visited and explored in a systematic and efficient manner.

Other algorithms, such as Dijkstra's, Bellman-Ford's, Kruskal's and Prim's, are designed to solve specific problems, such as the shortest path problem, the minimum spanning tree problem and the maximum flow problem.

Graph theory is a fascinating and evolving field with many open questions and exciting challenges. Among these challenges are the search for characterizations and structural properties for specific classes of graphs, the study of random graphs and stochastic processes on graphs, and the development of efficient methods and algorithms for dealing with graphs of large size and complexity. Graph theory is closely related to other areas of mathematics, such as algebra, topology, geometry and combinatorics, and benefits from their methods and results. In particular, graph algebra, which is the focus of our study, lies at the intersection of linear algebra and graph theory, and provides a powerful and unified framework for representing, manipulating, and analyzing graphs from an algebraic perspective. In this section, we will explore the key concepts of graph theory, such as graphs, vertices, edges, graph types and properties, as well as representations of graphs, such as adjacency lists, adjacency matrices, and incidence matrices. We will show how these concepts and representations are closely related to the notions and methods of linear algebra, and how they allow us to unveil and exploit the structure and properties of graphs in an elegant and efficient way. In sum, graph theory is an exciting and fertile field of mathematics, which offers a myriad of concepts, methods and applications to model, analyze and solve complex and interconnected problems.

(*)The Königsberg bridges have long been a famous mathematical problem that has attracted the interest of many mathematicians. This problem consists of determining whether it is possible to cross the seven bridges of the city of Königsberg once and only once, leaving and returning to the same place, without ever crossing two bridges at the same time. The inhabitants of the city searched for a long time for a solution to this problem, without success. It was finally Leonhard Euler who succeeded in solving it in 1735, by transforming the problem into a question of topology using a concept called "graphs". This solution marked the beginning of a new branch of mathematics, known as "graph theory", which has since been used to solve many other mathematical and practical problems.

The four lands A, B, C, D are separated by rivers and each edge, representing a bridge between two lands, is 7.

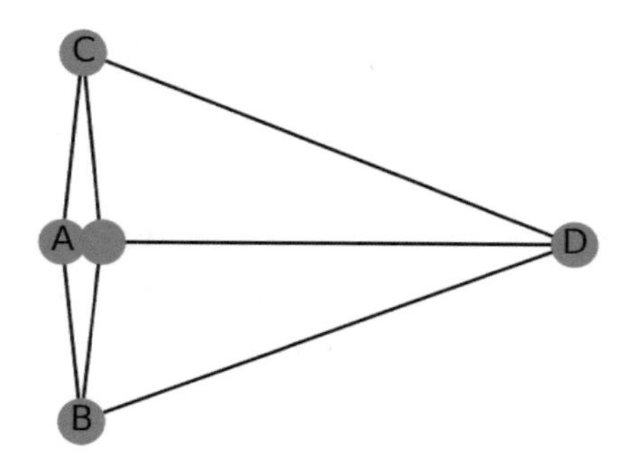

In order to understand the underlying structures and complex relationships behind relational data sets, we turn to the fundamental concept of graph theory: the graph itself. A graph is an abstract representation of a set of objects and the relationships between them. The objects are represented by nodes (or vertices), and the relationships between these objects are represented by edges (or links) that connect the nodes. Graphs can be classified into several categories, depending on the nature and constraints of their edges. Graphs can be undirected or directed. In an undirected graph, the edges have no direction, which means that the relations they represent are symmetrical. On the other hand, in a directed graph (also called a digraph), the edges have a direction, and the relations they represent are asymmetric. Graphs can also be weighted or unweighted. In a weighted graph, each edge is associated with a numerical value (or weight), which represents the strength, distance or cost of the relationship between the vertices it connects. Unweighted graphs, on the other hand, do not have weights on their edges, and all relations are considered equivalent. To illustrate the richness and diversity of graphs, let us consider some examples of graphs that frequently appear in practical applications.

Planar graphs are graphs that can be drawn on a plane without their edges intersecting, and they play an important role in geometry, topology, and graph algorithms. Trees are undirected, related graphs without cycles, which are ubiquitous in computer science, biology and game theory. To take the example of game theory, we will take the example of naval battle. In this game, each player places ships on a grid of given dimensions, and then fires shots to try to sink the opponent's ships. This grid can be represented as a planar graph, where each node represents a grid cell and the edges represent connections between adjacent cells (vertically or horizontally). If a cell is occupied by a ship, the corresponding node is considered occupied, while if a cell is empty, the corresponding node is considered free. The occupied vertices represent the locations of the ships, while the free nodes represent the possible locations for the players' moves. The edges model the rules of the game, which allow players to shoot only on adjacent squares and not to shoot twice on the same square. Thus, the corresponding graph represents the naval battle game grid, where each node is a square of the grid and the edges model the constraints of the game.

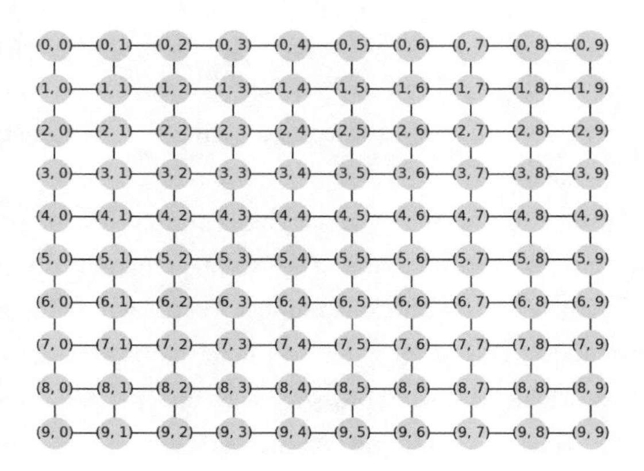

Bipartite graphs are graphs whose set of nodes can be partitioned into two subsets, such that all edges connect a node in one subset to a node in the other subset. Bipartite graphs are closely related to coupling, coloring, and partitioning problems, and they appear in various contexts, such as recommender systems, membership networks, and assignment problems. Bipartite graphs are often used to represent games where there are two types of players or objects that interact, such as two teams in a soccer game. In this case, the nodes of the first subset represent the players and the nodes of the second subset represent the teams. The edges of the bipartite graph represent connections between players and teams, for example if a player belongs to a given team.

Bipartite graphs allow modeling the rules of the game and the constraints associated with the membership of players in a given team, which can be useful for strategic planning and performance analysis. For example, the bipartite graph generated by the Python code can represent such a game, where the nodes of the first subset (A, B, C, and D) represent the players and the nodes of the second subset (1, 2, and 3) represent the teams. The edges connecting the nodes of the two subsets represent the membership of the players to a given team. For example, the edge between A and 1 represents the fact that player A belongs to team 1.

In our study of graphs, we will encounter many properties and invariants that characterize and distinguish graphs, such as the degree of a node (which is the number of edges incident to that node), the distance between two nodes(which is the minimum number of edges to travel from one node to the other), and the diameter of a graph (which is the largest distance between two nodes of the graph). These properties will allow us to quantify and compare graphs, and to better understand their structure and behavior.

A fundamental concept in the study of graphs is the notion of path, which is a sequence of nodes and edges connected in a consecutive manner. Paths play a central role in many problems and algorithms in graph theory, such as finding the shortest path between two nodes, detecting cycles and determining the connectedness of a graph. Connectedness is an important property of undirected graphs, which measures the "cohesion" of the graph and indicates whether each pair of nodes is connected by a path. A graph is said to be connected if, for each pair of nodes, there is a path connecting them. For example the graph A -- B -- C -- D -- E is connected contrary to A -- B -- C D -- E -- F which is not.

In the case of directed graphs, we speak instead of strong connectedness, which requires the existence of a directed path between each pair of nodes (e.g., A --> B --> C --> A). As we progress in our study of graphs, we will discover a world of amazing shapes and structures, where the simplicity of the basic concepts gives rise to a wide variety of problems, techniques and applications.

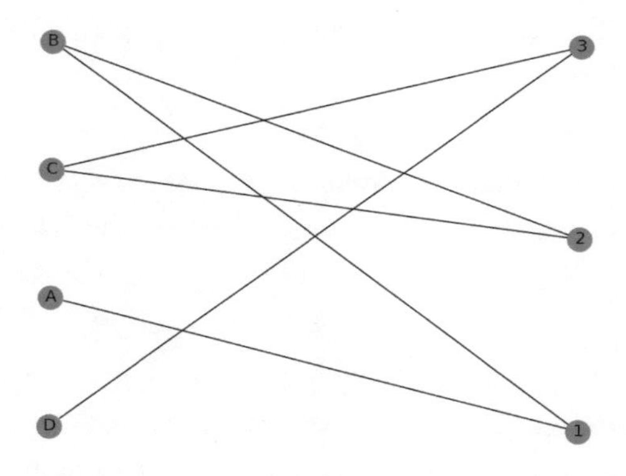

In this section, we will explore the diversity of graph types and their properties, to reveal the subtlety behind these concepts. We will look at various classifications and characteristics of graphs, which will help us to better understand their structure and behavior, and to identify features and patterns that can be exploited in relational data analysis and processing. First, we will discuss the distinction between directed and undirected graphs. A graph is said to be directed if each edge has a direction, i.e., it connects one node to another asymmetrically, whereas a graph is said to be undirected if each edge connects two nodes symmetrically. This fundamental difference has important implications for the structure and properties of graphs, and for the algorithms and techniques that can be used to analyze and manipulate them.

We will then examine various classifications of graphs based on their structure and properties, such as simple graphs (which do not have loops or multiple edges between the same nodes), regular graphs (in which all nodes have the same degree), planar graphs (which can be drawn without edges crossing each other), and bipartite graphs (whose nodes can be partitioned into two distinct sets, such that all edges connect a node in one to a node in the other). Each of these types of graphs has specific characteristics and properties that can be exploited to develop more efficient analysis methods and techniques. As part of our exploration of graph types, we will also be interested in various properties and invariants of graphs, such as the chromatic number (which is the minimum number of colors needed to color nodes so that two adjacent nodes have different colors), the number of spanning trees (which is the number of cycle-free subgraphs that contain all nodes) and the Euler characteristic (which is a measure of the "complexity" of a planar graph, based on the number of nodes, edges and faces).

These invariants play an important role in the characterization and classification of graphs, and allow us to better understand their structure and behavior. Finally, we will discuss some local properties and structures of graphs, such as cliques (which are complete subgraphs, i.e. subgraphs in which each pair of nodesis connected by an edge), cycles (which are closed paths with no repeated nodes, except for the first and the last) and stars (which are subgraphs consisting of a central node and its neighbors, with no other edges between them). The study of these structures allows us to better understand the topology and connectivity of graphs, as well as the patterns and relationships that may exist between their nodes and edges. The analysis of graph types and their properties is an essential step in mastering graph theory but also in developing efficient methods and techniques for the analysis and processing of relational data.

Graph representation is a fundamental issue in graph theory and graph algebra. A proper representation not only allows us to visualize and manipulate graphs, but also to apply algebraic and algorithmic methods to analyze and process them. In this section, we will examine different representations of graphs and discuss their advantages and disadvantages. The first representation we will consider is the diagram representation, which consists of drawing the nodes of the graph as points and the edges as lines or curves connecting these points.

This representation is particularly useful for visualizing the structure and topology of graphs, and for identifying patterns and relationships that may exist between their nodes and edges. However, the diagram representation also has drawbacks, including readability and ambiguity of the drawings, as well as difficulties in handling and processing large or structurally complex graphs. Another important representation of graphs is the adjacency list representation, which consists in describing a graph by the set of its nodes and the list of nodes adjacent to each of them. This representation is particularly adapted to hollow graphs, i.e. graphs whose number of edges is relatively small compared to the total number of possible pairs of nodes.

In a complete undirected graph, the total number of possible pairs of nodes is equal to n(n - 1)/2, where n is the number of nodes. Let's take an example with 5 nodes:

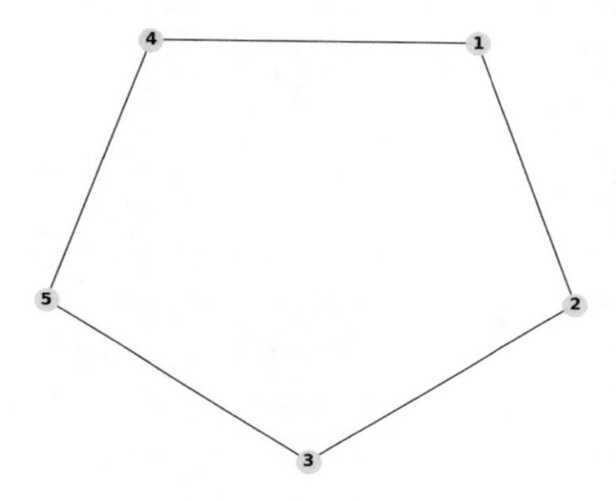

*In this graph, we have 5 nodes and 5 edges. The total number of possible pairs of nodes for a complete graph with 5 nodes is 5 *(5 - 1) / 2 = 10. Here, we have less than half of the maximum possible number of edges (5 out of 10), which makes it a hollow graph.*

Adjacency list representation allows quick and efficient access to the neighbors of a given node , which facilitates the implementation of graph traversal and exploration algorithms.

Matrix representation is another common method for representing graphs, which consists in associating a graph with a matrix whose elements reflect the structure and properties of the graph. Among the most commonly used matrices in graph algebra are the adjacency matrix, the incidence matrix and the Laplacian matrix.

Here is an example of the adjacency matrix for an undirected graph:

$$[A = \begin{bmatrix} 0 & 1 & 1 & 0 \\ 1 & 0 & 1 & 1 \\ 1 & 1 & 0 & 1 \\ 0 & 1 & 1 & 0 \end{bmatrix}]$$

Here is an example of the incidence matrix for an undirected graph:

$$[B = \begin{bmatrix} 1 & 1 & 0 & 0 \\ -1 & 0 & 1 & 0 \\ 0 & -1 & 1 & 1 \\ 0 & 0 & -1 & -1 \end{bmatrix}]$$

And here is an example of the Laplacian matrix for the same graph:

$$[L = \begin{bmatrix} 2 & -1 & -1 & 0 \\ -1 & 3 & -1 & -1 \\ -1 & -1 & 3 & -1 \\ 0 & -1 & -1 & 2 \end{bmatrix}]$$

As a reminder, the Laplacian matrix is a square matrix that can be associated with a graph, and is used in many areas of applied mathematics, such as graph theory, numerical analysis and physics. This matrix is often used to study properties of graphs, such as their connectivity and strength, as well as to solve partial differential equations on graphs. The Laplacian matrix L of an undirected graph G is defined as the difference between the degree matrix D and the adjacency matrix A of G. More precisely, the Laplacian matrix is defined as: L=D-A where D is the diagonal degree matrix of G, i.e., a diagonal matrix whose diagonal elements are the degrees of the nodes of the graph, and A is the adjacency matrix of the graph.

In the Laplacian matrix example I gave earlier, I considered a simple undirected graph with four nodes, and computed its Laplacian matrix. This matrix is a symmetric matrix, which means that $L_{ij} = L_{ji}$ *for all i,j, and is also a positive definite matrix, which means that its eigenvalues are all positive. These properties are important because they are used to solve partial differential equations on graphs, such as the heat equation or the Schrödinger equation, using the eigenvalues and eigenvectors of the Laplacian matrix.*

These matrices play a central role in the analysis and processing of graphs, allowing the application of algebraic techniques and methods to solve problems such as path finding, graph partitioning and graph optimization. Finally, the object and pointer representation is a more general method for representing graphs, which consists in describing a graph by a set of objects and pointers representing the nodes and edges, as well as the relations between them. This representation is particularly well suited to object-oriented programming languages, and it offers great flexibility and modularity for the implementation of graph processing algorithms and techniques.

It is important to note that no one representation is superior to the others, and the choice of the most appropriate representation will depend on the characteristics and requirements of the problem being solved, as well as the resources and constraints of the processing environment. For example, the diagram representation may be preferable for graph visualization and comprehension problems, while the adjacency list representation and the matrix representation may be more suitable for graph analysis and processing problems. We will also examine the relationships and transformations between the different representations, showing how one representation can be converted into another and how properties and results obtained in one representation can be transferred or generalized to other representations. This approach allows us to take advantage of the advantages and complementarities of the different representations, and to enrich our understanding and mastery of graph theory and graph algebra. By understanding the advantages and disadvantages of each representation, as well as the relationships and transformations between them, you will be better equipped to address the challenges and opportunities of graph research and application in various fields, such as social network analysis, recommender systems, bioinformatics and genomics, and many others.

Chapter 3: Graph algebra

1) Matrices associated with graphs

In this section, we will explore graph matrices, algebraic objects that allow for a formal and compact representation of graphs, as well as an efficient analysis and treatment of their properties and relationships. Graph matrices are a bridge between graph theory and linear algebra, uniting concepts and techniques from both fields to form a powerful and versatile framework for the study of graphs and their applications. There are several types of matrices associated with graphs, each with its own characteristics and application areas. In this section, we will focus on three particularly important and widely used types of matrices: the adjacency matrix, the incidence matrix and the Laplacian matrix.

We will introduce each of these matrices, describing their structure, construction and basic properties, and showing how they can be used to represent and analyze graphs and their elements, such as nodes, edges, paths and cycles. The adjacency matrix is a square matrix whose entries indicate whether two nodes are adjacent, that is, whether they are connected by an edge. The adjacency matrix captures the adjacency relationships between nodes and makes it easy to compute quantities such as the degree of nodes,, the number of triangles, and the number of paths of given length. The adjacency matrix is also used to define and solve problems such as graph coloring, clique detection, and finding isomorphic subgraphs.

The incidence matrix is a rectangular matrix whose entries indicate whether a node is incident to an edge, i.e., whether it is one of the ends of the edge. The incidence matrix captures the incidence relationships between nodes and edges and is used to calculate quantities such as the number of loops, the number of cycles, and the number of cuts. The incidence matrix is also used to define and solve problems such as cycle detection, connectivity determination, and construction of plungable graphs. The Laplacian matrix is a matrix derived from the adjacency matrix and the degree matrix, which measures the degree of nodes. The Laplacian matrix captures the neighborhood relationships between nodes and allows us to compute quantities such as conductance, effective strength, and the number of related components. The Laplacian matrix is also used to define and solve problems such as community detection, spectral partitioning and dynamic stability analysis.

In exploring graph matrices, we will strive to present the concepts and techniques in a clear and intuitive manner, with an emphasis on understanding the underlying ideas rather than mastering the technical details. We will show how matrices associated with graphs can be used to represent and analyze graphs in an elegant and efficient way, leveraging the tools and techniques of linear algebra to solve problems and answer questions about graphs and their properties. We will also examine the connections and interactions between the different types of matrices associated with graphs, showing how they complement and reinforce each other to provide a coherent and integrated overview of graphs and their characteristics. For example, we will see how the Laplacian matrix can be obtained from the adjacency matrix and the degree matrix, and how the spectral properties of the Laplacian matrix are related to the structural and topological properties of the underlying graph.

By approaching graph matrices with a rigorous yet accessible approach, we hope you will develop a deep appreciation for these fascinating mathematical objects and their central role in the study of graphs and their applications. We encourage you to deepen your knowledge of matrices associated with graphs and their properties, and to explore the many ramifications and extensions of these concepts in areas such as graph spectral analysis, graph geometry and topology, and machine learning on graphs. Finally, we will highlight the importance of graph-associated matrices in the processing of relational data and real-life problems. Graph matrices provide a formal and practical framework for representing, modeling, and analyzing complex and interconnected systems, such as social networks, biological systems, transportation infrastructures, and information systems. By mastering the concepts and techniques of graph matrices, you will be better equipped to address the challenges and opportunities in these areas and to contribute to the advancement of knowledge and technology in our interconnected world.

The adjacency matrix is one of the most basic and common representations of a graph.

In this section, we will examine this representation and its properties in detail, highlighting its usefulness and relevance in the study of graphs and their applications. Consider a graph G=(V, E), where V is the set of nodes and E is the set of edges. The adjacency matrix A of G is a square matrix of size |V| x |V|, where |V| is the number of nodes of the graph. The elements of the matrix A are defined as follows:

$$A_{ij} = \begin{cases} 1, & \text{if } (v_i, v_j) \in E \\ 0, & \text{otherwise} \end{cases}$$

In other words, the adjacency matrix simply encodes the presence or absence of edges between each pair of nodes in the graph. Edges are represented by 1's in the matrix, and the absence of edges is represented by 0's. Let's look at some important properties of the adjacency matrix. First, for an undirected graph, the adjacency matrix is always symmetric, that is $A_{ij} = A_{ji}$ for all i and j. This reflects the fact that, in an undirected graph, the edges are bidirectional and connect nodes in an undifferentiated way. For a directed graph, on the other hand, the adjacency matrix is generally not symmetric, because the edges have a direction and can connect nodes asymmetrically. Another interesting property of the adjacency matrix is its link with the degrees of the nodes. The degree of a node is the number of edges that are incident to it. For an undirected graph, the degree of a node v_i can be obtained by summing the elements of the i-th row (or the i-th column) of the adjacency matrix :

$$d(v_i) = \sum_{j=1}^{|V|} A_{ij}$$

Moreover, the adjacency matrix allows us to easily calculate the number of edges in the graph. For an undirected graph, the total number of edges is equal to half the sum of all the elements of the adjacency matrix:

$$|E| = \frac{1}{2}\sum_{i=1}^{|V|}\sum_{j=1}^{|V|} A_{ij}$$

For a directed graph, it is sufficient to sum all the elements of the adjacency matrix, without dividing by two:

$$|E| = \sum_{i=1}^{|V|}\sum_{j=1}^{|V|} A_{ij}$$

The adjacency matrix is also useful for studying paths and cycles in a graph. In particular, by raising the adjacency matrix to a certain power, one can obtain information about paths of different lengths in the graph. Consider the adjacency matrix A raised to the power k (i.e.,). The matrix product implies that the element (i, j) of corresponds to the number of paths of length k connecting node to node.

$$A_{ij}^{k} = \text{number of paths of length k between } v_i \text{ and} v_j$$

This property is particularly useful to detect the presence of cycles in a graph. A cycle is a closed path that returns to its starting point without repeating any edges. To detect the presence of cycles of length k, it is sufficient to check if the diagonal elements of A^k are non-zero. If $A_{ii}^k > 0$ for some i, it means that there is a cycle of length k passing through the node v_i.

The adjacency matrix also plays a central role in spectral graph analysis, which is a powerful technique for studying the structural and dynamic properties of graphs from their matrix representations. Spectral graph analysis relies on the study of the eigenvalues and eigenvectors of the adjacency matrix (and other associated matrices, such as the Laplacian matrix) to obtain information about global and local properties of the graph, such as connectivity, expansion, resilience and community detection.

In summary, the adjacency matrix is a fundamental and versatile representation of graphs that allows the study of many interesting properties and problems related to graphs. Its simplicity and clarity make it a valuable tool in the study of graphs and their applications in mathematics, computer science, physics and other fields.

To explore and understand graphs, we encounter another matrix representation, the incidence matrix, which offers us a complementary look at the structure and properties of graphs. While the adjacency matrix focuses on the relationships between pairs of nodes, the incidence matrix focuses on the relationships between nodes and edges, revealing more subtle and nuanced aspects of the topology of graphs. For an undirected graph G=(V, E) with n nodes and m edges, the incidence matrix is an n x m dimensional matrix B, where the element is defined as follows:

$$B_{ij} = \begin{cases} 1, & \text{if node } v_i \text{ is incident to edge } e_j \\ 0, & \text{otherwise} \end{cases}$$

In other words, each column of the incidence matrix corresponds to an edge, and the two non-zero elements of that column indicate the vertices that are connected by that edge. Note that for directed graphs, the incidence matrix is slightly different. In this case, $B_{ij} = 1$ if the node v_i is the head of the edge e_j, $B_{ij} = -1$ if the node v_i is the tail of the edge e_j, and $B_{ij} = 0$ otherwise. The incidence matrix offers a different perspective on graphs and allows one to study a range of problems and properties that are less obvious from the adjacency matrix. For example, the incidence matrix is particularly useful for studying cycles and cuts in a graph. As a reminder, a cycle is a closed path that returns to its starting point without repeating any edges, while a cut is a set of edges whose removal disconnects the graph. One of the remarkable results of graph theory, known as the cycle dimension theorem(*), establishes an intimate relationship between cycles and cuts in a graph.

(*) Known as the Krull-Schmidt theorem. It was discovered and developed by several mathematicians during the 20th century, notably by Wolfgang Krull and Otto Schmidt.

The cycle dimension theorem states that the vector space generated by the cycles of a graph is the orthogonal of the vector space generated by the cuts, when considering vector spaces over the two-element body (i.e., the coefficients are 0 or 1 and the addition is modulo 2). The proof of this theorem relies on the manipulation and analysis of the incidence matrix and its properties. The incidence matrix is also useful for studying traffic and flow problems in graphs, which have important applications in optimization, transportation, economics and other fields.

This graph shows an example of a simple graph with cycles and cuts. The blue solid edges represent cycles, while the red dashed edges represent cuts. In this example, the cycles are formed by nodes A, B, C and nodes C, D, E, while the cut is formed by the edge connecting nodes B and D.

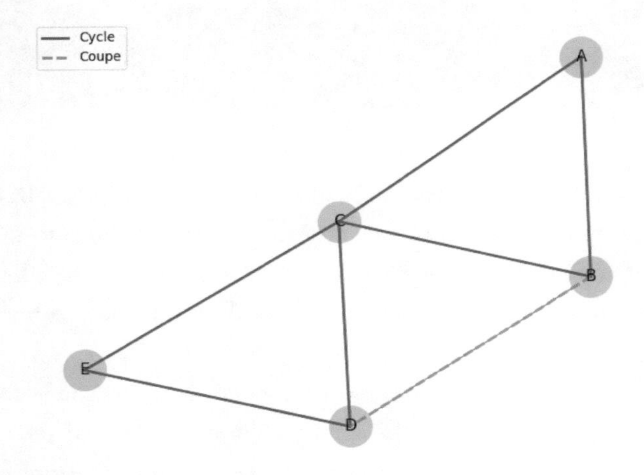

Traffic and flow algorithms exploit the structure and properties of the incidence matrix to solve problems such as determining the maximum traffic between two points in a network or minimizing transportation costs in a logistics system. Solutions to these problems can often be found using linear algebra and linear programming techniques, with the incidence matrix playing a central role in formulating and solving the constraints. For example, the Ford-Fulkerson algorithm is an iterative method for finding the maximum flow in a weighted (edge-capable) directed graph. It is based on the search for augmenting paths, i.e. paths that increase the flow between the source and the sink of the network. At each iteration, the algorithm finds an augmenting path and increases the flow along this path until there are no more augmenting paths. The sum of the flows along the final edges then gives the maximum flow.

Here is an example of pseudo-code of the Ford-Fulkerson algorithm:

```
Algorithm FordFulkerson(Graph G, Node source, Node sink):
    Initialize flow to 0 for all edges in the graph

    While there is a path P from source to sink in the residual graph:
        Let C be the capacity of the minimum-capacity edge in P

        For each edge e in P:
            If e is a forward edge in the original graph:
                Increase the flow of e by C
            Else If e is a backward edge in the original graph:
                Decrease the flow of e by C

    Return the sum of flows into the sink
```

The algorithm uses a breadth-first search (BFS) to find an augmenting path from the source to the sink, and updates the capacities of the edges along this path to increase the total flow of the graph. The main loop of the algorithm runs as long as an augmenting path can be found, and ultimately returns the computed maximum flow.

In addition, the incidence matrix can be used to analyze the connectivity and robustness of a graph. For example, the computation of the rank of the incidence matrix of a graph provides information on the connectivity of the vertices and the possible presence of isolated sub-graphs. Moreover, the incidence matrix allows us to study the vulnerabilities of the graph and to identify the edges or sets of edges whose deletion could have a disproportionate impact on the overall structure of the graph. Let's take the example of a road transportation network. Suppose we have a graph representing the roads in a region, where the nodes are the intersections and the edges represent the roads that connect these intersections. The incidence matrix of this graph can be used to study the connectivity of the intersections, i.e. whether all intersections are connected to each other by roads. To do this, one can calculate the rank of the incidence matrix, which gives the number of linearly independent rows in the matrix. If the rank is equal to the number of nodes, it means that all intersections are connected by roads. If the rank is less than the number of nodes, it indicates the presence of isolated subgraphs, i.e., areas of the network that are not connected to the rest of the network. In addition, the incidence matrix can be used to study the robustness of the transport network. By identifying edges or sets of edges whose removal could disproportionately impact the connectivity of the network, these edges can be strengthened to improve the robustness of the network. For example, if it is found that the deletion of a particular route splits the network into two disjoint parts, that route can be strengthened or an alternative route can be added to ensure network connectivity.

The incidence matrix also provides a means to characterize isomorphic graphs, that is, graphs that have the same structure but are labeled differently. Two graphs are isomorphic if and only if their incidence matrices are equivalent up to a permutation, meaning there exists a permutation of rows and columns that transforms one incidence matrix into the other. Although the graph isomorphism problem is notoriously difficult to solve in general, the incidence matrix provides a valuable tool for comparing and classifying graphs based on their structure. Lastly, the incidence matrix can be used to study the intersection and union of graphs. The intersection of two graphs is a graph whose edges are those belonging to both original graphs, while the union of two graphs is a graph whose edges are those belonging to either of the original graphs. By manipulating and combining the incidence matrices of the graphs in question, one can gain insights into the structure and properties of the resulting graphs. Overall, the incidence matrix offers a rich and nuanced perspective on graphs, revealing aspects of their structure and properties that are less evident from the adjacency matrix.

In this section, we will look at a key matrix related to graphs, called the Laplacian matrix. The Laplacian matrix is a powerful tool for studying the structural properties of a graph and for solving various problems in graph theory and its applications. It is particularly useful in the spectral analysis of graphs, where one examines the eigenvalues and eigenvectors of this matrix to obtain information about the structure of the underlying graph. The Laplacian matrix of an undirected graph G, denoted L(G), is defined as the difference between the degree matrix D(G) and the adjacency matrix A(G): $L(G) = D(G) - A(G)$

The degree D(G) matrix is a diagonal matrix of size n x n, where n is the number of nodes in the graph, and each diagonal entry d_{ii} is equal to the degree of the corresponding node v_i. The degree of a node is the number of incident edges at that node. The non-diagonal entries of D(G) are all zero. element L_{ij} of the Laplacian matrix is defined as follows:

$$L_{ij} = \begin{cases} d_i & \text{if } i = j \\ -1 & \text{if nodes } v_i \text{ and } v_j \text{ are adjacent} \\ 0 & \text{otherwise} \end{cases}$$

where d_i is the degree of the node v_i. The Laplacian matrix exhibits several remarkable properties that make it useful for graph analysis. First, it is symmetric, which implies that all of its eigenvalues are real. Moreover, the sum of the elements of each row (or column) of L(G) is equal to zero, which means that the constant vector $(1, 1, \ldots, 1)^T$ is an eigenvector of L(G) with eigenvalue zero. Also, the multiplicity of the eigenvalue zero in the Laplacian matrix is equal to the number of connected components of the graph. This can be used to determine whether a graph is connected or not, i.e. whether all nodes in the graph are reachable to each other by a sequence of edges. Another important property of the Laplacian matrix is its connection with the number of spanning trees of a graph. A spanning tree is an acyclic subgraph that contains all the nodes of the graph. The number of spanning trees of a graph can be calculated using the determinant of the modified Laplacian matrix, where any row and column are removed.

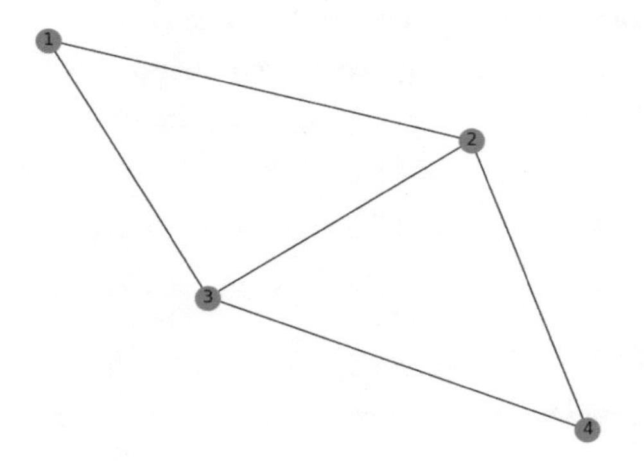

This graph is an undirected graph with 4 nodes and 5 edges. It represents a situation where there is a connection between every pair of nodes, except for one pair of nodes which is disconnected. The Laplacian matrix of this graph can be used to calculate the number of spanning trees, which is equal to the determinant of the modified Laplacian matrix.

Spectral analysis of graphs using the Laplacian matrix can also be used to solve graph partitioning and combinatorial optimization problems. For example, the spectral partitioning method uses the eigenvalues and eigenvectors of the Laplacian matrix to divide the nodes of a graph into distinct groups. This technique is used to identify communities in social networks or to solve load balancing problems in distributed systems.

Also, the Laplacian matrix is closely related to diffusion and propagation processes on graphs. Solutions of partial differential equations on graphs, such as the heat equation or the Schrödinger equation, can be expressed in terms of the eigenvalues and eigenvectors of the Laplacian matrix. This has applications in various fields, such as image analysis, modeling of sensor networks and the dynamics of biological networks. In sum, the Laplacian matrix is an essential tool for studying the structural and spectral properties of graphs. It offers a powerful way to understand and analyze graphs from various algebraic perspectives. Thanks to its many properties and applications, the Laplacian matrix plays a central role in graph theory, graph algebra and related fields. In the following sections, we will explore in more detail the algebraic operations on graphs and their implications on the structure of graphs. We will also look at how the various matrices associated with graphs, such as adjacency matrix, incidence matrix, and Laplacian matrix, can be used to solve practical problems in various application areas, such as analysis social networks and recommendation systems.

2) Algebraic operations on graphs

The study of graphs is closely linked to the algebraic operations associated with them. These operations allow us to combine and manipulate the graphs in various ways, revealing interesting relationships and structures. In this section, we will present the main algebraic operations on graphs and discuss their properties and their consequences on the structure of graphs. We will begin our exploration by studying the simplest operations on graphs, such as sum and product, which allow us to construct new graphs from existing graphs. We will study the properties of these operations and see how they can be used to reveal interesting and complex structures in graphs.

Then, we will look at more advanced operations, such as the power of a graph and the calculation of characteristic polynomials. These operations, although more difficult to grasp, have profound implications for graph theory and can be used to solve complex problems and to study the internal structure of graphs. The power of a graph is an operation that consists of taking a graph and "multiplying" it by itself a certain number of times. This operation is closely linked to the notion of paths in graphs and makes it possible to determine, for example, the number of paths of length k between two given nodes. By studying the powers of a graph, we can thus obtain information about the connectivity and the overall structure of the graph. The calculation of characteristic polynomials is an even more advanced operation, which involves the study of the eigenvalues and eigenvectors associated with the matrices defined on the graphs. Characteristic polynomials are powerful tools that allow us to extract information about the structure of graphs and to solve complex problems such as community detection and spectral analysis.

Algebraic operations on graphs are an elegant and natural way to study how graphs can be combined and manipulated. In this section, we are interested in two fundamental operations: the sum and the product of graphs. Let's start with the sum of graphs. Let $G_1 = (V_1, E_1)$and $G_2 = (V_2, E_2)$ two graphs. The sum of the graphs G_1 and G_2, noted $G_1 + G_2$, is a new graph whose node set is the union of the node sets of G_1 and G_2, and whose set of edges is the union of the sets of edges of G_1 and G_2.

More formally, we have: $G_1 + G_2 = (V_1 \cup V_2, E_1 \cup E_2)$

It is important to note that in this definition we assume that the nodes V_1 and V_2 are disjoint, that is, they do not share any nodes. If not, one can always uniquely label the nodes before doing the sum. The sum of graphs can be used to describe situations where two distinct graphs are combined without interaction between them, for example when merging two independent networks. It is interesting to note that the sum of graphs preserves some properties of the original graphs, such as their connectivity, their density and their number of cycles. However, it is not always very informative about the relationships between graphs G_1 and G_2, because it does not take into account any interactions between them. To further investigate the relationships between two graphs, we can use the product of graphs. There are several ways to define the product of graphs, each highlighting different aspects of the relationships between the graphs. We will focus here on two types of graph products: the Cartesian product and the tensor product.

The Cartesian product of two graphs $G_1 = (V_1, E_1)$ and $G_2 = (V_2, E_2)$, noted $G_1 \Box G_2$, is a graph whose node set is the Cartesian product of the node sets of G_1 and G_2, and whose edges connect the pairs of nodes (u_1, u_2) and (v_1, v_2)if and only if $u_1 = v_1$ and $(u_2, v_2) \in E_2$, or $u_2 = v_2$ and$(u_1, v_1) \in E_1$.

More formally, we have:

$G_1 \Box G_2 = (V_1 \times V_2, ((u_1, u_2), (v_1, v_2)) \mid (u_1 = v_1\ et (u_2, v_2) \in E_2)\text{ ou }(u_2 = v_2\text{ et }(u_1, v_1) \in E_1))$

The Cartesian product makes it possible to explore the relations between the graphs by constructing a new graph where each node represents a pair of nodes of the original graphs. The edges of the Cartesian product are determined by the edges of the graphs $G_1$and G_2, thus reflecting the relationships between these graphs. The Cartesian product is commutative, that is, $G_1 \square G_2 = G_2 \square G_1$, and it preserves some properties of the original graphs, such as their connectivity and chromatic number.

The tensor product of two graphs $G_1 = (V_1, E_1)2 = G_2 \square G_1$ and $G_2 = (V_2, E_2)$ noted $G_1 \otimes G_2$, is a graph whose node set is also the Cartesian product of the node sets of G_1 et G_2, and whose edges connect the pairs of nodes (u_1, u_2) and (v_1, v_2) if and only if $(u_1, v_1) \in E_1$ and $(u_2, v_2) \in E_2$.

More formally, we have:

$$G_1 \otimes G_2 = (V_1 \times V_2, ((u_1, u_2), (v_1, v_2)) \mid (u_1, v_1) \in E_1 \text{ et } (u_2, v_2) \in E_2)$$

The tensor product creates a new graph which is a kind of "fusion" of the original graphs, in that the edges of the tensor product are determined by the edges of the graphs G_1 and G_2 simultaneously. Unlike the Cartesian product, the tensor product is not commutative in general. The tensor product can be used to study combinatorial problems and to construct graphs with particular properties, such as strongly regular graphs or graphs with a large chromatic number.

Consider two simple graphs as concrete examples to illustrate the tensor product. Let G be a graph with two connected nodes (a graph consisting of a single edge) and H a cycle with four nodes (square).

```
G:  1 -- 2


H:  1 -- 4
    |    |
    2 -- 3
```

The tensor product of G and H, denoted G⊗H, is determined as follows:

For each pair of nodes (u, v) in G and (x, y) in H, there is an edge between (u, x) and (v, y) in G⊗H if and only if u is connected to v in G and x is connected to y in H.

Here is the correct representation of the tensor product G⊗H:

```
(1, 1) -- (1, 4) -- (2, 3) -- (2, 2)
   |         |         |         |
(1, 2) -- (1, 3) -- (2, 4) -- (2, 1)
```

In this example, the tensor product G⊗H is a graph with 8 nodes and 8 edges. The edges of the graph G ⊗ H reflect the connections in both G and H.

The tensor product can be used to construct graphs with particular properties, as mentioned before. For example, if G and H were strongly regular graphs, then G⊗H would also be a strongly regular graph. Moreover, the tensor product can be used to study combinatorial problems and to construct graphs with a large chromatic number, as in the case of Ramanujan graphs.

Here is an example of a regular Ramanujan graph:

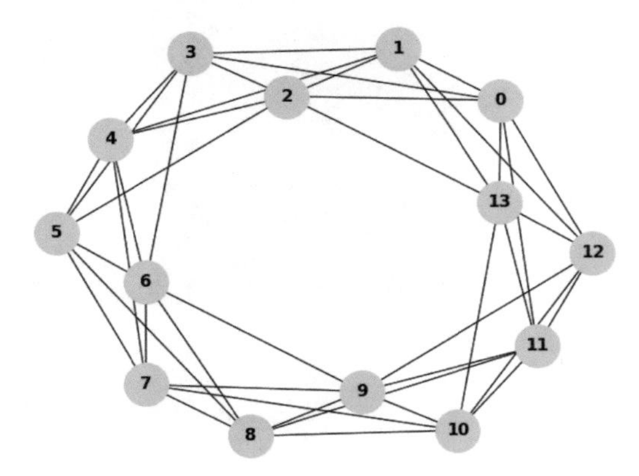

In summary, the sum and the product of graphs are fundamental algebraic operations which make it possible to study the relations between graphs and to construct new graphs with interesting properties. Graph sum combines two graphs with no interaction between them, while graph products, such as Cartesian product and tensor product, explore the relationships between graphs by constructing a new graph whose nodes and edges are determined by the original graphs. These operations provide a rich and flexible framework for the study of graphs and their properties, and they are a valuable tool for graph algebra.

The power of a graph is another interesting algebraic operation that can be used to study the properties of a graph and the relationships between its nodes.
The power of a graph $G = (V, E)$of order n, noted G^k, is defined as the graph whose nodes are the same as those of G and whose two nodes are adjacent if and only if they are at a distance of k in the graph G. More precisely, let $d(u, v)$) the distance between nodes u and v in G, i.e. the length of the shortest path connecting u and v. So, G^k is a graph such that:

$$G^k = (V, (u, v) \mid d(u, v) = k)$$

To illustrate this notion, let us take the example of a graph G formed by a cycle of four nodes , that is to say a square. For G^2 ,the nodes are the same as those of G, and two nodes are adjacent in G^2 if they are at a distance of 2 in G. In this case, G^2 is a complete graph, i.e. a graph where each pair of nodes is connected by an edge. It is important to note that the power of a graph is not the same as the Cartesian product or the tensor product of graphs. The power of a graph is defined only in terms of the structure of a single graph, while graph products involve the combination of two separate graphs.

Studying the powers of a graph can reveal interesting information about the structure and properties of the graph. For example, if G^k is a complete graph for some k, then this means that the graph G has the property that every pair of nodes can be connected by a path of length k. Moreover, the power of a graph can be used to solve combinatorial problems and to construct graphs with particular properties, such as strongly regular graphs or graphs with a large chromatic number. Let us take the example of a graph G formed by a cycle of five nodes, that is to say a pentagon. If we calculate G^2, we get a graph where every pair of non-adjacent nodes in G is now adjacent in G^2. In this case, G^2 is also a complete graph, with each pair of nodes connected by an edge.

Another interesting example is the Petersen graph, a regular undirected graph of order 10 and degree 3. The Petersen graph is a strongly regular graph and has many interesting properties. If we calculate the power G^2 from the Petersen graph, we get a graph each vertex of which is adjacent to all other nodes except itself and its neighbors in the original graph. G^2 of the Petersen graph is therefore a graph with structural properties different from those of the original graph. Power analysis of a graph can also be useful for studying the spectral properties of a graph. In particular, the eigenvalues of the adjacency matrix of a graph G and those of the adjacency matrix of its powers G^k are linked. Indeed, if λ is an eigenvalue of the adjacency matrix A of G, then λ^k is an eigenvalue of the adjacency matrix of G^k. This relation makes it possible to deduce information on the eigenvalues of G from those of its powers and vice versa.

Here is the simple Petersen graph:

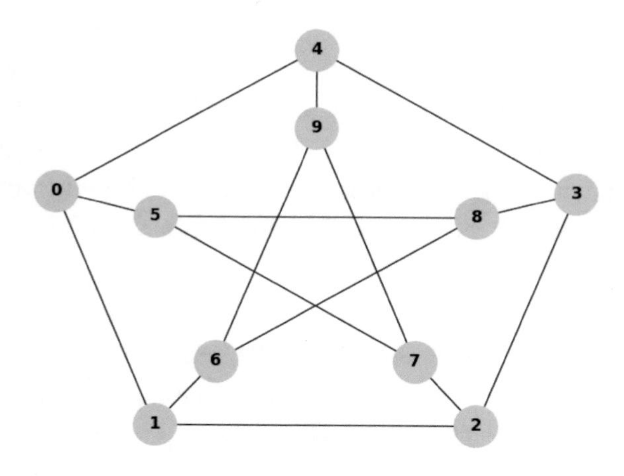

Here is the Petersen graph to the power of 2:

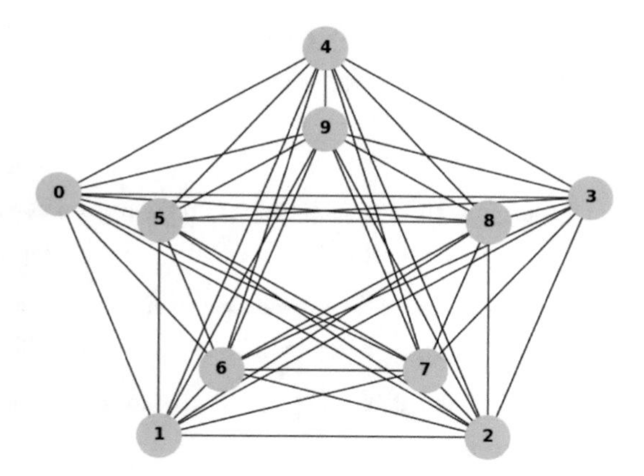

Moreover, the power of a graph can be used to study optimization and path finding problems. For example, if one seeks to determine the shortest path between two nodes in a weighted graph, one can calculate the power of the graph corresponding to the weights of the edges, then find the path having the minimum weight in this power of the graph. Similarly, the power of a graph can be used to solve node and edge cover problems, where one seeks to select a minimal subset of nodes or edges such that each node or edge of the graph is covered by at least one element of the subset.

In sum, the power of a graph is an algebraic operation that allows the exploration of many interesting properties of graphs and offers applications in various fields, such as optimization, path finding and spectral analysis. As with other algebraic operations on graphs, the power of a graph is closely related to the structure of the graph and can provide valuable information about the relationships between its nodes and the overall properties of the graph. By mastering this operation, researchers and practitioners can develop a deeper understanding of graphs and their potential applications in various contexts.

This pseudo-algorithm presents how to use the power of a graph to find the shortest path between two nodes and solve node and edge covering problems.

Pseudo-algorithm:

```
CalculatePowerGraph(G, n) function:
   If n == 1, return G

   G_power_n = CopyGraph(G)

   For each pair of vertices i and j in G:
      If the distance between i and j in G is equal to n:
         Add an edge between i and j in G_power_n

   Return G_power_n
```

Function PlusShortPathWeight(G, weight, node_start, node_end):
G_weight = CalculatePowerGraph(G, weight)

Apply a shortest path algorithm (e.g. Dijkstra) on G_weight with node_start and node_end

Return shortest path found

MinimumCoverNode(G) function:
Find a set of nodes S_minimal such that each edge of G is incident to at least one node of S_minimal

Return S_minimal

MinimumEdgeCover(G) function:
Find a set of edges E_minimal such that each node of G is incident to at least one edge of E_minimal

Return E_minimal

The characteristic polynomials and the spectrum of a graph are two closely related notions which make it possible to extract information about the structure and properties of a graph from its associated matrices. These concepts have applications in various fields such as network analysis, theoretical chemistry and the study of the spectral properties of graphs. The characteristic polynomial of a graph is a polynomial defined from the adjacency matrix of the graph. Let A be the adjacency matrix of a graph G with n nodes. The characteristic polynomial of G, denoted $P_G(x)$,is defined by : $P_G(x) = det(xI - A)$, where x is an independent variable, I is the identity matrix of size n and det() denotes the determinant of a matrix.

The characteristic polynomial is a polynomial function of degree n whose coefficients are determined by the entries of the adjacency matrix. The spectrum of a graph, on the other hand, is the set of its eigenvalues, that is to say the solutions of the characteristic polynomial. The eigenvalues of a graph provide important information about the structure of the graph and its properties. For example, they can be used to determine if a graph is bipartite, to calculate the spectral distance between two graphs, or to study the coupling properties of graphs. It is also interesting to note that the characteristic polynomial and the spectrum of a graph are invariant under isomorphism. In other words, if two graphs are isomorphic, their characteristic polynomials and their spectra are identical. This property is useful for graph isomorphism detection and graph classification. Consider a simple example to illustrate these concepts. Let G be a graph with three vertices and three edges, forming a cycle of length three. The adjacency matrix A of G is given by:

$$A = \begin{pmatrix} 0 & 1 & 1 \\ 1 & 0 & 1 \\ 1 & 1 & 0 \end{pmatrix}$$

The characteristic polynomial of G is :

$$P_G(x) = det(xI - A) = x^3 - 3x + 2$$

and the eigenvalues of G are the solutions of this equation: λ_1 = 2, λ_2 = -1 and λ_3 = -1. The spectrum of G is therefore {2, -1, -1}. The characteristic polynomial and the spectrum of a graph also have applications in the study of optimization and partitioning problems on graphs. For example, the method of spectral partitioning, which aims to divide a graph into subgraphs of balanced size and low interconnection, relies on the use of eigenvalues and eigenvectors associated with the Laplacian matrix of the graph.

In summary, the characteristic polynomial and the spectrum of a graph are powerful mathematical tools that allow inferring valuable information about the structure and properties of graphs from their associated matrices. They are used in various fields of application, such as network analysis, theoretical chemistry and the study of the spectral properties of graphs. Moreover, these concepts are invariant under isomorphism, which makes them particularly useful for the detection of isomorphisms of graphs and the classification of graphs. It should also be mentioned that research on characteristic polynomials and the spectrum of graphs is an active and growing field. Researchers continue to explore new properties and applications of these concepts, as well as to develop new techniques to efficiently compute characteristic polynomials and graph spectra, especially for large and complex graphs. Furthermore, understanding the relationships between characteristic polynomials, graph spectra, and other algebraic objects associated with graphs, such as Laplacian and incidence matrices, can also pave the way for significant advances in the study graphs and associated optimization problems.

In sum, characteristic polynomials and graph spectra are essential tools for graph algebra. They allow not only to extract information on the structure and properties of graphs, but also to address optimization and partitioning problems. Continued research in this area promises to bring new knowledge and applications for the analysis and manipulation of graphs in various scientific fields.

Chapter 4: Algorithms and techniques for processing relational data

1) Graph exploration

In this chapter, we are interested in graph exploration, an exciting branch of graph algebra that aims to study and understand the underlying structures of these mathematical objects. Graph mining is a rich and varied field, which encompasses a large number of methods and techniques for analyzing and solving problems in various fields such as operations research, computer science and social networks, to name a few. only a few. Before we dive into the details of algorithms and mining methods, it is essential to take a moment to reflect on the importance of the study of graphs in modern science.

Indeed, graphs are ubiquitous in many fields, ranging from the modeling of interactions between individuals in a social network, to the representation of chemical bonds in a molecule. The ability to understand, manipulate and analyze these mathematical objects is therefore of crucial importance for solving complex problems in a wide range of disciplines. Graph mining is a task that can be approached in several ways, depending on the objectives and constraints specific to each problem. For example, one can seek to determine the best way to traverse a graph, by visiting each vertex once and minimizing the total distance traversed. In another case, one may be interested in the detection of subsets of nodes having particular properties, such as the formation of communities or distinct groups. In all cases, the goal is to gain a deep understanding of the structure of the graph and to apply this knowledge to solve practical problems. This chapter is organized in such a way as to provide the reader with a thorough and rigorous introduction to the main techniques of graph exploration.

We start by describing depth-first and breadth-first search algorithms, which are fundamental tools for exploring graphs and discovering their structure. These methods, while seemingly simple, are extremely powerful and can solve many crawling problems in an efficient and elegant way. Next, we look at the Dijkstra and

Bellman-Ford algorithms, which are classic methods for solving the shortest path problem in weighted graphs. These algorithms, which were developed decades ago, are still widely used today due to their efficiency and robustness against various types of graphs. We present the key ideas behind these methods and show how they can be effectively implemented to solve graph mining problems.

Finally, we conclude this chapter by discussing some future perspectives and challenges in the field of graph mining. Although much progress has been made over the past few years, there is still much to discover and understand about how graphs can be effectively explored and analyzed. Among the open questions that deserve particular attention are the complexity of graph mining algorithms, the discovery of even faster and more efficient methods for solving complex problems, and the application of these techniques to new and emerging fields. It is also interesting to note that, alongside theoretical advances, graph mining has benefited from considerable advances in the field of information technology. Modern computers are now able to process massive data sets and perform complex calculations in record time, allowing researchers to study graphs of unprecedented size and complexity.

Additionally, the emergence of new technologies, such as parallel and distributed computing, offers exciting opportunities to push the boundaries of graph mining even further. In this context, it is essential that researchers and practitioners work hand in hand to develop new methods and techniques that will address the challenges of graph mining in an increasingly connected and complex world. This not only requires a deep understanding of underlying mathematical and algorithmic concepts, but also an ability to apply them in creative and innovative ways to solve real-world problems. In sum, this chapter aims to provide the reader with a comprehensive and rigorous overview of the state of the art in graph mining, while highlighting the challenges and opportunities that arise in this exciting field.

The notion of connectedness constitutes a fundamental element in the study of graphs, because it makes it possible to apprehend the global structure of a graph and to deduce certain interesting properties from it. In this section, we will define the notion of connectedness and examine how it relates to the notion of connected components, as well as the practical implications of these concepts in graph analysis.

Definition (Connectivity) : An undirected graph G=(V, E) is said to be connected if, for any pair of nodes u and v belonging to V, there is a path connecting u and v. In other words, it is possible to get from one node to any other node in the graph by following a sequence of edges.

The notion of connectedness is closely related to that of connected components, which are subsets of the graph having a certain particular structure. Intuitively, a connected component is a "piece" of the graph in which all nodes are connected to each other, and which is isolated from the rest of the graph in terms of connections.

Definition (Related component): A connected component of an undirected graph G=(V, E) is a subgraph H=(V', E') of G such that H is connected and V' is a maximal subset of V having this property. That is, a connected component is a connected subgraph of G which cannot be extended to a larger connected subgraph by adding nodes or edges of G.

Example: Consider the following graph G, composed of 10 nodes and 9 edges:

G = (V, E), where V = v1, ..., v10} and E = {e1, ..., e9} with:
e1 = (v1, v2), e2 = (v2, v3), e3 = (v3, v4), e4 = (v1, v4), e5 = (v5, v6), e6 = (v6, v7), e7 = (v5, v7), e8 = (v8, v9), e9 = (v9, v10)

In this graph, we can identify three related components:
C1 = (V1, E1), where V1 = {v1, v2, v3, v4} and E1 = {e1, e2, e3, e4}
C2 = (V2, E2), where V2 = {v5, v6, v7} and E2 = {e5, e6, e7}
C3 = (V3, E3), where V3 = {v8, v9, v10} and E3 = {e8, e9}

It is important to note that the connected components of a graph form a partition of the set of nodes, i.e. each node belongs to one and only one connected component. Moreover, the connectedness of a graph is directly related to the number of its connected components. A graph is connected if and only if it has only one connected component.

Property : Let G=(V, E) be an undirected graph. Then G is connected if and only if there is a spanning tree of G.

The previous property allows us to deduce the existence of spanning trees for connected graphs. A spanning tree is a subgraph of G which is a tree (i.e. without cycle) and which contains all the nodes of G. This notion is particularly useful in various applications, such as the construction of communication networks or solving combinatorial optimization problems.

In the case of directed graphs, the notion of connectedness is slightly more complex, because it depends on the direction of the edges. A distinction is then made between strong and weak connectivity.

Definition (Strong and Weak Connectivity): A directed graph G=(V, E) is said to be weakly connected if, ignoring the direction of the edges, the underlying undirected graph is connected. G is said to be strongly connected if, for any pair of nodes u and v belonging to V, there is a directed path connecting u to v and a directed path connecting v to u.

Weak and strong connectivity are interesting properties for analyzing the structure of directed graphs, in particular for studying social networks, transport networks, or biological interaction networks.

To conclude, connectedness and connected components are central concepts in graph analysis, allowing to better understand the global structure of a graph and to deduce interesting properties. The distinction between weak and strong connectivity for directed graphs offers additional perspectives for the study of these complex structures.

In this section, we discuss two important algorithms for solving the shortest path problem in weighted graphs, namely Dijkstra's algorithm and Bellman-Ford's algorithm. Each of these algorithms has its own advantages and disadvantages, and understanding them in depth is essential to choosing the best approach for each situation. Dijkstra's algorithm is an efficient method for finding the shortest path between a source node and all other nodes in an undirected weighted graph with non-negative weights on the edges. The central idea of this algorithm is to use a greedy approach to gradually build the shortest path from the source to each node. At each step, the algorithm chooses the unvisited node with the minimum distance to the source, updates the distances of the neighbors of this node and marks the node as visited. The algorithm ends when all nodes have been visited or when the minimum distance to the target node is found.

This graph illustrates Dijkstra's algorithm applied to an undirected weighted graph. Nodes are represented by blue circles with their labels, and edges are represented by black lines with their weights. The shortest paths from the source node (0) to all other nodes are represented by red lines.

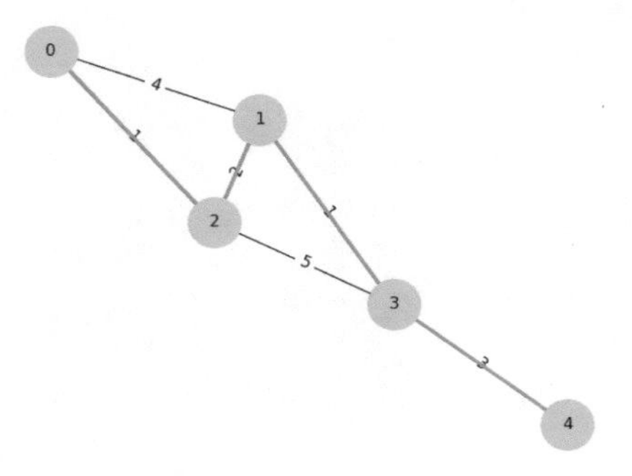

Consider the following example to illustrate Dijkstra's algorithm. Consider a graph G with nodes {A, B, C, D, E} and the following weighted edges: (A, B, 10), (A, C, 3), (B, C, 1), (B , D, 2), (C, D, 8), (C, E, 2), (D, E, 7), (E, A, 6). We want to find the shortest path from A to E. Applying Dijkstra's algorithm, we get the following steps:

1. Initialize the distances of all node to infinity, except for A which is 0.
2. Select A as the current node and update the distances of neighbors B (10), C (3) and E (6).
3. Mark A as visited and choose the unvisited node with the smallest distance, i.e. C.
4. Update the distances of C's neighbors: B (4), D (11), and E (5).
5. Mark C as visited and choose the unvisited node with the smallest distance, i.e. E.
6. Mark E as visited. Since E is the target node, the algorithm terminates.
7. The shortest path from A to E is A → C → E with a total distance of 5.

The Bellman-Ford algorithm is another method for finding the shortest path in a weighted graph. Unlike Dijkstra's algorithm, it can handle negative weights on edges and detect negative cycles. The Bellman-Ford algorithm works by relaxing all edges of the graph a certain number of times (usually equal to the number of nodes minus one). A relaxation of an edge consists in updating the distance between two nodes if the passage by this edge makes it possible to reduce this distance. The Bellman-Ford algorithm also detects negative cycles by checking, after the last relaxation, whether further relaxation further reduces the distance between two nodes. If so, it means there is a negative cycle in the graph.

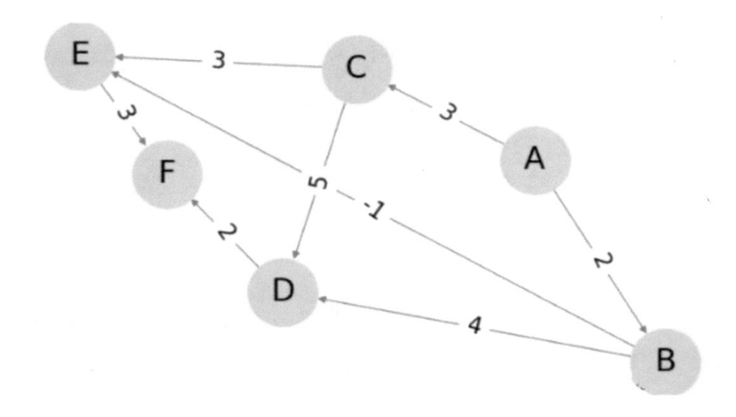

This graph illustrates the Bellman-Ford algorithm applied to a weighted directed graph. Nodes are represented by blue circles with their labels, and edges are represented by gray arrows with their weights. The shortest paths from the source node to all other nodes are represented by red arrows. This algorithm can handle negative weights on edges and detect negative cycles in the graph.

Let's take an example to illustrate the Bellman-Ford algorithm. Consider a graph G with nodes {A, B, C, D} and the following weighted edges: (A, B, 5), (A, C, 4), (B, C, -2), (B, D, 3), (C, D, 2), (D, B, 1). We want to find the shortest path from A to D. Applying the Bellman-Ford algorithm, we get the following steps:

- Initialize the distances of all nodes to infinity, except for A which is 0.
- Perform edge relaxations for each node(n-1 times, or 3 in our example):
  - 1st iteration: A → B (5), A → C (4), B → C (3), B → D (8), C → D (6), D → B (7).
  - 2nd iteration: A → B (5), A → C (4), B → C (3), B → D (8), C → D (6), D → B (7).
  - 3rd iteration: A → B (5), A → C (4), B → C (3), B → D (8), C → D (6), D → B (7).
- Perform additional relaxation to detect negative cycles.

In this example, no distance is reduced, so there is no negative cycle. The shortest path from A to D is A → C → D with a total distance of 6. In summary, Dijkstra's algorithm is generally faster than Bellman-Ford's algorithm, but it cannot handle graphs with negative weights on the edges. The Bellman-Ford algorithm, on the other hand, is more flexible, as it can handle negative weights and detect negative cycles. The selection of the appropriate algorithm will therefore depend on the characteristics and constraints of the problem to be solved.

2) Graph partitioning

Graph partitioning is a fundamental problem in graph theory, which consists in dividing a graph into sub-graphs, depending on some characteristics or constraints.

Spectral methods are techniques based on linear algebra and the analysis of the eigenvalues of matrices associated with graphs, such as the adjacency matrix or the Laplacian matrix. These methods have been widely studied and applied in various fields, such as social network analysis, biology, chemistry, and physics. To understand the essence of spectral methods, we must first discuss a few fundamental concepts. The spectrum of a graph is the set of its eigenvalues, sorted in ascending order. The smallest eigenvalue of a Laplacian matrix is always zero, and its associated eigenvector is called the fundamental eigenvector.

The other eigenvalues and eigenvectors are called upper eigenvalues and upper eigenvectors. The spectrum structure of a graph gives important information about the properties of the graph, such as its connectivity, regularity, and bipartiteness. The spectral partitioning of a graph is carried out using the information contained in the eigenvectors associated with the smallest non-zero eigenvalues of the Laplacian matrix. The simplest method is to use the second smallest eigenvector, called the Fiedler eigenvector. The spectral partitioning algorithm can be described as follows:

- Calculate the Laplacian matrix L of the graph G. Find the Fiedler eigenvector (the second smallest eigenvector) of L.
- Use the components of Fiedler's eigenvector to determine the partitioning of the graph.

In the case of an unweighted graph, the sign of the components of the Fiedler eigenvector can be used to partition the nodes into two sets. If the component of the eigenvector corresponding to a noed is positive, the node is assigned to one set, otherwise it is assigned to the other set. For weighted graphs, a similar approach can be taken using a threshold determined by the edge weights.

This graph illustrates the spectral partitioning applied to the famous "Zachary's Karate Club", an unweighted and undirected graph. Nodes are represented by circles with their labels and edges are represented by gray lines connecting the nodes. The graph is split into two separate subgraphs using Fiedler's eigenvector: one colored red and the other blue. Red nodes are those with positive Fiedler eigenvector components, while blue nodes have negative components.

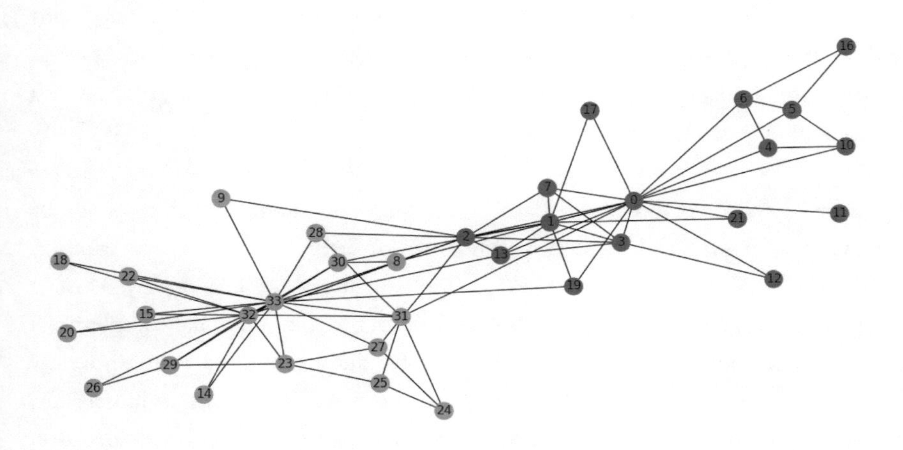

Let's take a simple example to illustrate the spectral method of partitioning graphs. Consider a graph G with nodes {A, B, C, D, E, F} and the following unweighted edges: (A, B), (A, C), (B, C), (C, D) , (D, E), (D, F), (E, F). By applying the spectral method, we obtain the following steps:

- Calculate the Laplacian matrix L of G.
- Find the Fiedler eigenvector of L, which is approximately (-0.37, -0.37, -0.22, 0.22, 0.37, 0.37) in this case. 3.
- Using the components of Fiedler's eigenvector, partition the nodes into two sets. Nodes A, B and C have negative components, while nodes D, E and F have positive components.

Thus, the graph is partitioned into two subgraphs: {A, B, C} and {D, E, F}. It should be noted that spectral partitioning can also be extended to k-part partitioning problems, using the first k nonzero eigenvectors of the Laplacian matrix. These eigenvectors are then combined to form a matrix, the rows of which are then clustered using clustering methods, such as k-means clustering. Spectral methods have several advantages over other graph partitioning methods. First, they are based on fundamental algebraic properties of graphs and provide information about the overall structure of the graph. Moreover, they are relatively simple to implement and can be easily adapted to different types of graphs, such as weighted graphs, directed graphs and multipart graphs. Finally, the spectral methods have a good performance in terms of partitioning quality, in particular for graphs presenting a community structure or a weak coupling between the sub-graphs.

However, spectral methods also have limitations. For example, they may not be suitable for graphs with complex structures or irregular connection patterns. Moreover, spectral partitioning is based on approximations and may not give an exact solution to the graph partitioning problem. Finally, the computational complexity(*) of spectral methods can be high for large graphs, due to the need to compute the eigenvalues and eigenvectors of the Laplacian matrix. Despite these limitations, spectral methods remain a valuable tool for graph partitioning and relational data analysis. Future research could focus on improving spectral algorithms, exploring new graph representations, and integrating spectral methods with other relational data processing techniques.

Stream methods are another important approach for graph partitioning, based on the optimization of streams in networks. These methods model the graphs as transport networks, where the edges represent communication channels between the nodes and the weights of the edges represent transport capacities. The objective of graph partitioning is then to divide the graph into subgraphs in such a way as to minimize the sum of the capacities of the cut edges, that is, the edges that connect the different subgraphs.

(*) Number of operations and amount of resources needed (time and space) to execute an algorithm depending on the size of the input, here the size of the graph.

A classical flow algorithm for graph partitioning is the Stoer-Wagner algorithm, which solves the minimum cut problem. This algorithm works by repeating the following steps until the graph is reduced to a single node:

- Select an arbitrary node in the graph and apply the maximum flow algorithm to find the minimum cut between this node and all other nodes in the graph.
- Merge the two nodes of the minimum cut found in step 1 into a single node and update the edge weights accordingly.
- Repeat steps 1 and 2 for the reduced graph.

The Stoer-Wagner algorithm guarantees to find the global minimal cut of the graph, which corresponds to an optimal partitioning of the graph into two subgraphs. However, this algorithm cannot be applied directly to partition a graph into more than two subgraphs. In this case, extensions and variants of the Stoer-Wagner algorithm can be used, such as recursive partitioning methods or multi-level partitioning methods.

Let's take a concrete example to illustrate the Stoer-Wagner algorithm. Suppose we have a social network consisting of 5 people (A, B, C, D, E) and the connections (friendships) between them. The weights on the edges represent the strength of their relationships, where higher values indicate stronger relationships. The graph is represented as follows:

We want to partition this social network into two groups, so that the sum of the strengths of the relationships between the groups is minimized. Let's apply the Stoer-Wagner algorithm:

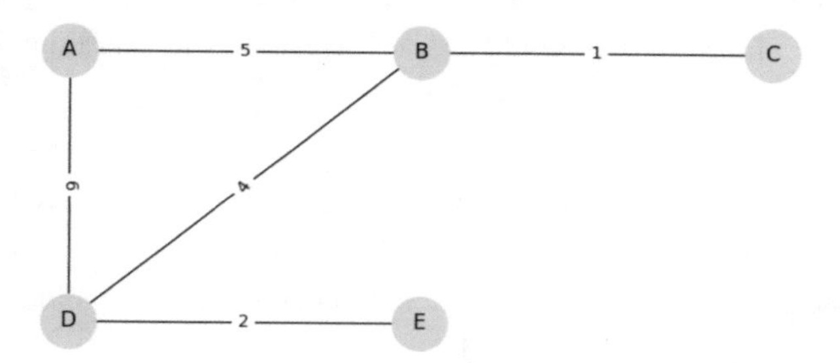

- Let's choose an arbitrary node, say A. Let's find the minimum cut between A and all other nodes using a maximum flow algorithm. In this case, the minimum cut is obtained by separating {A, B, D} and {C, E} with a total weight of 3 (edge B-C).
- Let's merge the nodes of the minimum cut found in step 1, i.e. B and C, into a single node (B-C) and update the edge weights accordingly:

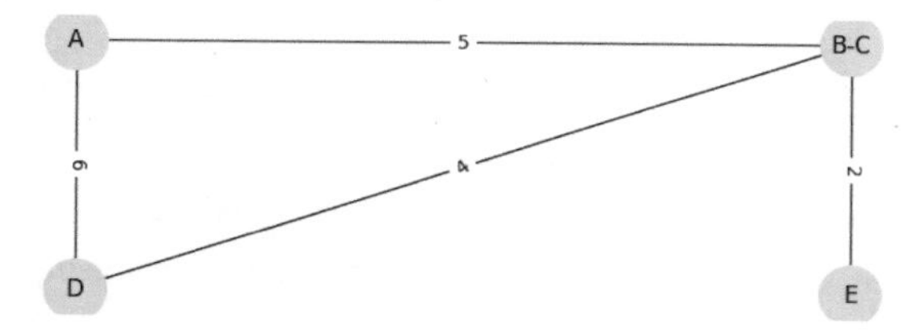

- Let's repeat steps 1 and 2 for the reduced graph. This time the graph is reduced to a single node (A-B-C-D-E), and the global minimum cut is the one found previously: {A, B, D} and {C, E}.

Thus, the Stoer-Wagner algorithm partitions the social network into two groups: {A, B, D} and {C, E}, minimizing the sum of the strengths of the relationships between the groups.

3) Graph optimization

A fundamental aspect of graph theory is the study of couplings, which are subsets of disjoint edges, that is, not sharing common nodes. Couplings are useful for solving many optimization problems in areas such as resource allocation, network planning and design. In this section, we will examine the properties of couplings, as well as some methods for determining them.

This graph illustrates an example of coupling in an undirected graph. Node are represented by blue circles with their labels, and edges are represented by gray lines. The selected pairing, consisting of edges (A, B), (C, D) and (E, F), is highlighted in red, showing a set of disjoint edges that do not share common nodes.

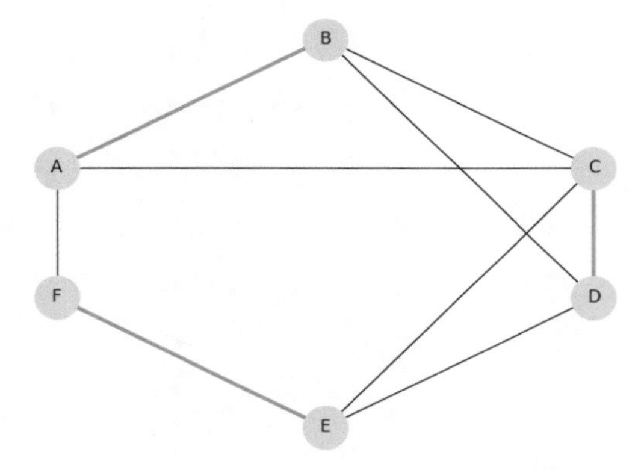

Let's start by defining some essential terms relating to couplings. A matching M in a graph G is a set of disjoint edges, i.e. each node is incident to at most one edge of M. A node is said to be coupled (or saturated) if v is incident to an edge of M, and uncoupled (or unsaturated) otherwise. A matching is maximal if it is not possible to add an edge to M without deleting another one. A matching is of maximum cardinality if no other matching has more edges. Finally, a matching is perfect if all the nodes are matched.

It is important to note that the maximum cardinality pairing is not necessarily unique. For example, a complete graph with an even number of nodes may have several different maximum cardinality pairings. However, all these pairings have the same size, i.e. they have the same number of edges. One of the most famous algorithms for finding a maximum cardinality matching is the Hungarian algorithm, also known as the Kuhn-Munkres(*) algorithm. It is a combinatorial optimization algorithm, based on the node labeling method and the path augmentation method, which mainly works on weighted bipartite graphs. This algorithm finds a perfect matching of maximum weight, where the weight is the sum of the weights of the edges in the matching. Let's take an example to illustrate the application of the Hungarian algorithm. Consider a weighted bipartite graph G = (A ∪ B, E), where A = {a1, a2, a3} and B = {b1, b2, b3} are the two sets of nodes, and the edge weights are as follows :

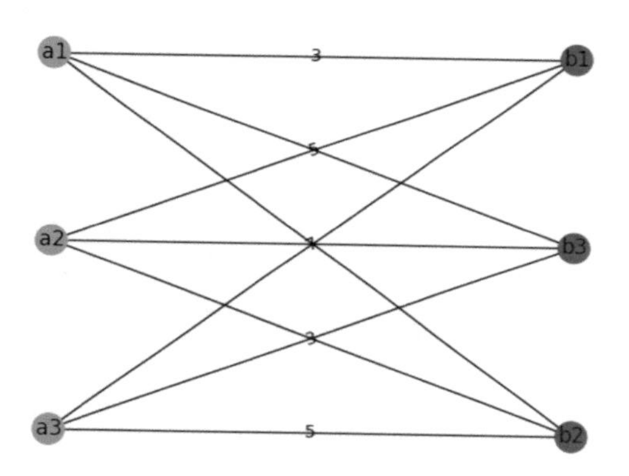

(*) The Kuhn-Munkres algorithm, also known as the Hungarian algorithm, was independently developed by Harold Kuhn and James Munkres in 1955 and 1957, respectively. It is widely used to solve assignment problems, such as assigning tasks to workers or matching people in a market. It has a time complexity of , where n is the number of nodes in the bipartite graph.

The Hungarian algorithm finds the perfect matching of maximum weight {(a1, b3), (a2, b1), (a3, b2)} with a total weight of 15. This matching is optimal because there is no other perfect matching with greater weight.

The shortest path problem is a classic and fundamental problem in graph theory, which consists in determining the shortest path between two nodes of a graph. There are many applications for this problem, such as navigation systems, route planning, and communication networks. To illustrate this problem, consider a weighted graph G = (V, E) with positive or zero weights on the edges, where V is the set of nodes and E is the set of edges. The weight of a path is the sum of the weights of all the edges that compose it. The goal is to find the path of minimum weight between two given nodes, often called source and destination.

To illustrate these concepts, consider a weighted graph G with nodes{A, B, C, D, E} and the following edges: {(A, B, 4), (A, C, 2), (B, C , 1), (B, D, 5), (C, D, 8), (C, E, 10), (D, E, 2)}. Edge weights are listed third in each tuple.

Suppose we wanted to find the shortest path between A and E. Using Dijkstra's algorithm, we would do the following:

- Initialize the distances: $A = 0, B = \infty, C = \infty, D = \infty, E = \infty$
- Select node A (distance 0) and update the distances of its neighbors:
  - A -> B : 0 + 4 = 4
  - A -> C : 0 + 2 = 2

- Select node C (distance 2) and update the distances of its neighbors:
  - C -> B : 2 + 1 = 3 (distance shorter than 4)
  - C -> D : 2 + 8 = 10
  - C -> E : 2 + 10 = 12
- Select node B (distance 3) and update the distances of its neighbors:
  - B -> D : 3 + 5 = 8 (distance shorter than 10)
- Select node D (distance 8) and update the distances of its neighbors:
  - D -> E : 8 + 2 = 10 (distance shorter than 12)

Node E is reached, the algorithm terminates. The shortest path between A and E is therefore A -> C -> B -> D -> E, with a total weight of 10.

Using the Bellman-Ford algorithm, we would get the same result after several iterations of relaxations. It is important to note that Bellman-Ford's algorithm is generally slower than Dijkstra's algorithm, but it can handle graphs with negative edge weights.

In conclusion, the shortest path problem is an essential problem in graph theory with many practical applications. Dijkstra's algorithm and Bellman-Ford's algorithm are two classic approaches to solving this problem, each having its advantages and disadvantages. The choice of algorithm depends on the characteristics of the graph and the requirements of the application.

The traveling salesman problem is a classic combinatorial optimization problem that consists of determining the shortest path to visit a set of nodes of a weighted graph only once and then return to the starting point. This problem is often presented as a representative of NP-hard problems (*), for which there is no known polynomial-time efficient solution.

To illustrate the problem, consider a graph G with 4 nodes {A, B, C, D} and the following weights for each pair of nodes: w(A, B) = 10, w(A, C) = 15, w (A, D) = 20, w(B, C) = 35, w(B, D) = 25, w(C, D) = 30.

The traveling salesman's problem is to find the shortest circuit through each node exactly once and back to the starting point. In this example, the shortest circuit is A -> B -> C -> D -> A, with a total weight of 75.

(*) This means that it belongs to a class of problems for which there is no known algorithm capable of finding an optimal solution in polynomial time, i.e. a time which increases most rapidly as a function of the size of the problem entry.

It is important to note that the traveling salesman problem is very different from the shortest path problem. While the shortest path problem can be solved efficiently using the Dijkstra and Bellman-Ford algorithms, the traveling salesman problem is NP-hard and there is no known polynomial algorithm to solve it. .

Nevertheless, there are several approaches to dealing with the traveling salesman problem, including exact methods and heuristics. Exact methods, such as linear programming, dynamic programming, and implicit enumeration, guarantee the optimal solution but can be very time-consuming and memory-intensive for large problems.

Heuristics are approximate methods that aim to find a solution close to the optimum in a reasonable time. Some commonly used heuristics for the traveling salesman problem are:

1. Nearest neighbor heuristic: selects the node closest to the current node as the next step.
2. The minimal spanning tree heuristic: constructs a minimal spanning tree of the graph and visits the nodes of this tree in depth.
3. Metaheuristics: global search methods, such as genetic algorithms, tabu search, and simulated annealing, that explore the solution space by performing combinations of local moves.
4. Ant colony algorithms: simulate the deposition and evaporation of pheromones on the edges of the graph, guiding artificial ants to build solutions.
5. Guided local search: explores the space of solutions by performing local improvements and using guidance mechanisms to escape local optima.

These heuristics do not always guarantee the optimal solution, but they can provide an acceptable solution in a reasonable time, especially for large problems. To illustrate the application of the nearest neighbor heuristic to the traveling salesman problem, consider the previously mentioned graph G. Starting from node A, the heuristic will first select B as the next step, then D, then C, and finally return to A. The resulting circuit is A -> C -> B -> D -> E -> A, with a total weight of 65.

In conclusion, the traveling salesman problem is a complex combinatorial optimization problem that cannot be solved efficiently in polynomial time. However, various exact and heuristic methods have been developed to deal with this problem. The choice of method will depend on the specific constraints and requirements of the problem to be solved.

Chapter 5: Practical Applications of Graph Algebra

1) Social media analysis

The modern world is deeply marked by the exponential growth of social networks. From Facebook to Twitter, via LinkedIn and Instagram, people are constantly connected to each other, weaving links and interactions that form complex structures. Analyzing these structures requires appropriate tools and techniques, which can be found in the study of graphs and linear algebra. In this section, we will examine the main concepts and methods used in social network analysis, with an emphasis on the practical applications of graph algebra.

First, it is important to understand how social networks can be represented as graphs. The nodes of the graph correspond to individuals or entities (for example, users, companies or organizations), while the edges represent the relationships between them (friendship, collaboration, affiliation, etc.). Graphs used in social network analysis can be directed or undirected, weighted or unweighted, depending on the nature of the relationships and the data available. For example, an undirected graph may be used to represent symmetric friendship relationships, while a directed graph may be needed to model asymmetric relationships, such as Twitter followings or hierarchical relationships within a company.

One of the key concepts in social network analysis is centrality, which measures the importance of a node in the graph. There are several ways to define centrality, depending on the specific aspects of the network that one wishes to highlight. The main measures of centrality are:

<u>Degree centrality:</u> based on the number of incident edges at a node, it measures the importance of an individual in terms of direct connections. In a social network, a node with a high degree centrality can be considered a key player or an influencer.

<u>Proximity centrality:</u> calculated as the inverse of the sum of the distances between a node and all the other nodes of the graph, it measures the accessibility of an individual

to the entire network. A node with high proximity centrality is well positioned to spread information or quickly reach other members of the network.

Betweenness centrality: based on the number of shortest paths passing through a node, it measures the importance of an individual as a bridge or intermediary between other members of the network. A node with high betweenness centrality is crucial for maintaining network cohesion and can play a key role in controlling information flow.

These centrality measures can be computed using algebraic techniques, including matrix multiplication, matrix inversion, and diagonalization. For example, the proximity centrality can be determined from the distance matrix, which contains the shortest distances between each pair of nodes. Similarly, betweenness centrality can be computed using shortest path search algorithms, such as the Dijkstra or Bellman-Ford algorithm, presented in Section 4.1.2.

In addition to centrality indicators, the study of social networks also requires the detection of communities or groups of individuals who have stronger links between themselves than with the rest of the network. To accomplish this task, one can resort to graph partitioning techniques, such as spectral methods and flow methods, which are discussed in Section 4.2. For example, by performing the spectral decomposition of the Laplacian matrix of the graph, it is possible to reveal community structures by taking advantage of the properties of the eigenvectors associated with the smallest eigenvalues. This approach highlights groups of individuals with closer ties to each other. Other methods, such as the Louvain algorithm or modularity optimization, are also used to detect communities more efficiently and adaptable to large networks. These approaches allow the network to be divided into distinct subgroups, identifying the ties that connect individuals with stronger relationships or common interests. Thus, the analysis of social networks is enriched thanks to the detection and understanding of the community structures that emerge from them.

In this example graph, imagine that the nodes represent users of a social network, and the edges represent friendships between them. The two communities could then be interpreted as groups of friends with closer ties to each other than to other network users.

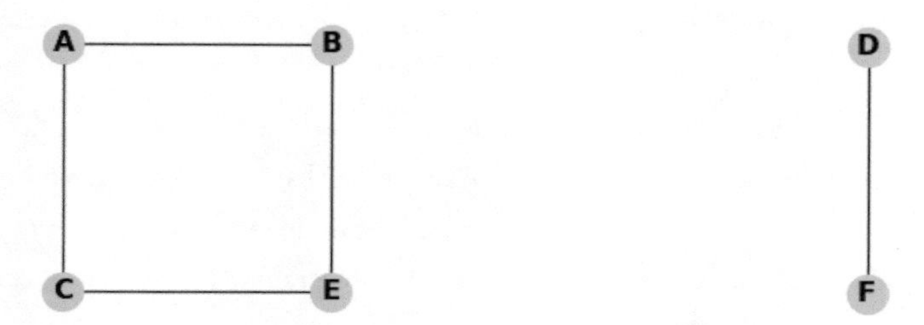

Community 1 (A, B, C, E): Alice, Bob, Carol and Eve are a group of friends who know each other well and interact with each other frequently. They share common interests and participate in social events together.

Community 2 (D, F): On the other side, David and Frank are two friends who have a close relationship, but are not connected to other members of the network. They could be work colleagues or members of a sports club, sharing specific interests.

The graph thus illustrates how graph partitioning methods can be used to detect communities or groups of individuals with stronger ties to each other than to the rest of the network, providing a better understanding of community structures within networks. social.

Social network analysis has many practical applications, ranging from marketing and advertising to surveillance and epidemic prevention. By understanding the mechanisms that govern the interactions and dynamics of social networks, researchers and practitioners can design more effective strategies to influence the behavior of individuals, promote the dissemination of information or products, and identify opinion leaders. or key actors in different contexts.

In this section, we will explore the concepts of centrality measurement and community detection in more detail, presenting concrete examples and case studies drawn from various fields of application. We will also discuss the challenges and limitations of social media analysis, with a focus on issues of privacy, ethics, and bias in data. Finally, we will look at future prospects and technological developments that could shape the future of social network analysis, such as machine learning on graphs, advanced visualization techniques, and multimodal data integration.

Centrality measures are essential tools to characterize the importance of the nodes of a graph, especially in the context of social networks. There are several measures of centrality, each emphasizing a different aspect of node importance. The main measures of centrality are degree centrality, proximity centrality, betweenness centrality and Eigenvector centrality (*).

Degree centrality is the simplest measure and is based on the number of edges incident to a node. The more neighbors a node has, the more central it is considered. For an undirected graph, the degree centrality of a node v is simply equal to its degree d(v). In a directed graph, we distinguish the centrality of indegree and the centrality of outdegree, corresponding respectively to the number of edges entering and leaving the node.

Proximity centrality measures the accessibility of a node relative to other nodes in the graph. It is defined as the inverse of the sum of the shortest distances between the node in question and all the other nodes. The higher the proximity centrality of a node, the closer it is to other nodes, and therefore likely to facilitate the dissemination of information or to influence other individuals in the network.

(*) The term "Eigenvector" is of German origin, "Eigen" means "own" or "characteristic" in German, and "vector" refers to a vector. Thus, "Eigenvector" can be roughly translated as "eigenvector" or "characteristic vector".

Betweenness centrality, on the other hand, captures the importance of a node as a "bridge" or crossing point between other nodes. It is calculated by counting the number of shortest paths passing through the considered node, and by normalizing this quantity by the total number of shortest paths in the graph. A node with high betweenness centrality plays a crucial role in the overall network connectivity and can serve as a checkpoint to control or monitor information flows. Eigenvector centrality is a more complex measure that takes into account not only the number of neighbors of a node, but also their respective importance. It is based on the spectral decomposition of the adjacency matrix of the graph and assigns to each nodea value of centrality proportional to the sum of the centralities of its neighbors. This measure makes it possible to detect influential nodes even if they are not directly connected to a large number of other nodes, but rather to nodes that are themselves important.

Take, for example, a social network where individuals are represented by nodes and friendships by edges. Degree centrality can help identify the most popular individuals with the greatest number of friends. Proximity centrality makes it possible to identify individuals who are well connected to the entire network and can therefore quickly spread information or rumours. The centrality of betweenness highlights individuals who serve as a bridge between different social groups, playing a key role in the flow of information between these groups. Finally, the centrality of Eigenvector allows to detect influential individuals who are connected to other influential people, even if they do not necessarily have a large number of friends.

In this graph, we have 8 individuals, represented by the nodes from 'A' to 'H'. Friendship relationships are represented by the links between these nodes. The network structure is rather centralized around individual 'D', who has the largest number of direct friends (4 in total), which results in a high degree centrality. Therefore, 'D' is probably the most popular individual in this network. Individual 'D' also has a high proximity centrality, which means that he is well connected to the whole network and can therefore quickly spread information or rumors. Regarding the centrality of betweenness, the individual 'D' also plays an important role by serving as a bridge between different social groups, in particular the groups 'A-B-C' and 'E-F-G-H'. Finally, even though the centrality of Eigenvector is not visually obvious in the graph, it would be calculated to assess influential individuals connected to other influential people. This description highlights the key properties of the social network represented by this graph, based on the measures of centrality discussed previously.

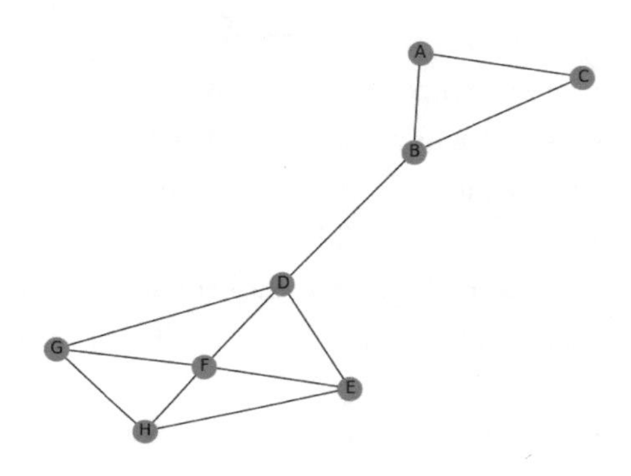

It is important to note that these different measures of centrality are not always correlated, and that a node can be central according to one measure while being marginal according to another. Therefore, it is often useful to combine multiple metrics to get a more complete view of the importance of nodes in a network. Moreover, these centrality measures can be adapted to take into account additional attributes of nodes or edges, such as weights representing the intensity of relations or geographic distances. For example, in a transportation network, proximity centrality can be calculated using travel times rather than straight line distances between nodes.

To illustrate the application of centrality measures, consider a fictitious social network comprising nine individuals (A, B, C, D, E, F, G, H and I) and the friendship relationships between them. By calculating the different centrality measures for each individual, one can obtain interesting information on the structure of the network and the role of each individual. For example, one might find that individual A has a high degree centrality, indicating that he is very popular, while individual B has a high betweenness centrality, suggesting that he plays a key role in the connectivity between different social groups.

The degree centrality for each node is the number of links it has with others. For example, if 'A' has a high degree centrality, it means he has a lot of friendships, so he is very popular. Betweenness centrality for each node measures how many times a node is on the shortest path between two other nodes. If 'B' has a high betweenness centrality, it means that it is often on the shortest path between two other people, suggesting that it plays a key role in connectivity between different social groups.

In sum, centrality measures are powerful tools for studying social networks and understanding the roles and positions of individuals within these networks. By combining these measures with techniques from graph algebra and relational data analysis, it is possible to extract valuable insights into the mechanisms underlying the formation and evolution of social networks, as well as the dynamics of influence and dissemination of information within these networks.

In this part, we are interested in the detection of communities in social networks, a central issue of graph analysis. Communities are groups of nodes that are closely related to each other and share common properties or attributes. The identification of these communities makes it possible to better understand the structure and the social dynamics of the networks, to optimize the dissemination of information and to detect unusual behavior.

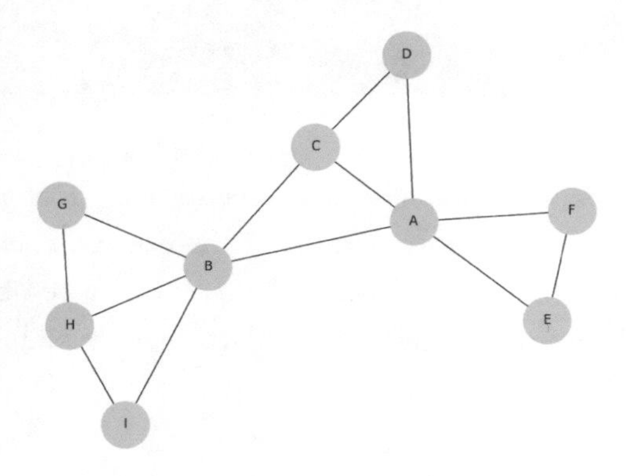

A common method to detect communities is the maximization of modularity. Modularity measures the density of links within communities relative to links between communities. A high modularity value indicates that the communities are well defined. The Leuven algorithm, for example, is an effective method for optimizing modularity in large social networks.

Let's take a concrete example: let's consider a network of friendships in a school, where the pupils are the nodes and the friendship relations, the edges. By applying the Louvain algorithm, we can identify several communities corresponding to different classes and groups of friends. Knowledge of these communities would, for example, make it possible to target specific messages or events to each group. Here is a pseudo-code explaining the principle of the Louvain algorithm:

```
1. Initialization:
    - Assign each node its own community
      (i.e, if you have N nodes, you have N communities)

2. Modularity Optimization Phase:
    - For each node, calculate the modularity gain if the node
      is removed from its community and placed in the community of each of its neighbors
    - If the modularity gain is positive, move the node into the community with the
    highest gain
    - Repeat these two steps for all nodes until the modularity cannot be improved
     (i.e., the community for each node is optimal)

3. Community update phase:
    - Create a new "view" of the network by replacing each community with a single node.
      The weights of the links between the new nodes are equal to the sum of the weights
      of the links between the nodes in the corresponding communities.
    - The links within each community result in self-directed loops on the new nodes.

4. Repeat phases 2 and 3 until the modularity cannot be improved anymore.

5. The final result is a partition of the network into communities.
```

Another approach to detect communities is the spectral analysis of graphs, which is based on the eigenvalues and the eigenvectors of the matrices associated with the graphs, such as the Laplacian matrix. Spectral techniques make it possible to divide the graph into sub-graphs in such a way as to minimize inter-community links while maintaining strong internal consistency.

To return to our school example. Using spectral analysis, we could detect groups of students based on their participation in clubs or sports. These groups could be distinct from the communities identified by the Louvain algorithm, thus offering a complementary perspective on the social structure of the school.

Finally, community detection can also rely on label propagation methods, such as the label propagation algorithm of Raghavan, Albert and Kumara (*). This algorithm assigns a unique label to each node and updates these labels based on neighbor labels, until most nodes have the most frequent label in their neighborhood.

(*) Raghavan, U. N., Albert, R., & Kumara, S. (2007). Near linear time algorithm to detect community structures in large-scale networks. Physical Review E, 76(3), 036106.

In our school example, the tag propagation algorithm could help detect groups of students based on their affinities, location, or extracurricular activities. These groups could also differ from previously identified communities, providing yet another view into the social dynamics of the school.

Suppose a small friendship network in a school with 5 students: Alice, Bob, Charlie, David and Elsa. Friends are represented as follows: Alice is friends with Bob and Charlie. Bob is also friends with Charlie. David and Elsa are friends with each other. Leuven's algorithm can be applied to our school friendship network example as follows:

```
1. Initialization:
    Each student (Alice, Bob, Charlie, David, Elsa) is assigned to their own community.
2. Modularity Optimization Phase:
    For each student, calculate the modularity gain if this student is removed from their
    community and placed in the community of each of their friends.
    If the modularity gain is positive, move the student to the community with the highest
    gain.
    Repeat these two steps until the modularity can no longer be improved.
3. Community Update Phase:
    Create a new "view" of the network by replacing each community with a single node.
    Internal links within each community are translated into self-directed loops on the
    new nodes.
5. Repeat phases 2 and 3 until the modularity can no longer be improved.
6. The final result is a partition of the network into communities.
```

For example, Alice, Bob and Charlie could be identified as one community, while David and Elsa could be identified as another community. These results could be used to better understand the social structure of the school. It is interesting to use the Leuven algorithm and other community detection methods to identify groups of friends in a school network, as it can help to understand social dynamics and identify students who could benefit from greater social integration. Additionally, it can help target specific messages or events to each group, improving communication and engagement within the school.

It is important to note that different community detection methods can produce different and complementary results. Thus, it may make sense to use multiple methods to gain a more nuanced understanding of the structure of social networks. Within the framework of the analysis of social networks, the detection of communities is a crucial issue for many applications. For example, it can help researchers identify groups of individuals who may share political views, business interests, or cultural preferences. In addition, companies may use this information to target their advertisements or personalize their offers based on identified communities.

In summary, community detection is an essential task in social network analysis, which helps to better understand the structure and dynamics of networks. The methods presented here, such as modularity maximization, spectral analysis, and label propagation, offer varied and complementary approaches to identify groups of closely related individuals. In the context of our increasingly connected society, these techniques are crucial to fully exploit the potential of relational data and better understand human behaviors.

2) Recommender systems

In a world increasingly rich in information and content, recommendation systems have become essential to help users navigate and discover the elements that suit them best. These systems are used in many fields, such as music or film streaming platforms, online sales sites or even social networks. The main objective of these systems is to anticipate and suggest the elements that best meet the needs, preferences and interests of the user.

Graph algebra, with its ability to model and analyze complex relationships between entities, provides a powerful framework for developing and improving recommender systems. In this section, we will explore two major approaches in the design of these systems: collaborative filtering and content-based models. We will show how concepts and techniques from graph algebra can be applied to improve the quality of recommendations and better understand user preferences.

Let's start with a concrete example to illustrate the importance of recommender systems. Let's say you've recently discovered an interest in "house music" and want to find other titles similar to those you've already enjoyed. You could browse playlists or compilations online, but the sheer number of tracks available makes this a tedious and time-consuming task. An effective recommendation system would then be able to offer you a list of relevant titles based on your previous listenings and your preferences, saving you time and giving you a personalized experience.

Collaborative filtering, the first approach we will discuss, is based on the idea that users with similar preferences in the past are likely to have common interests in the future. This approach can be divided into two subcategories: user-based collaborative filtering and item-based collaborative filtering. In the first case, the system identifies other users with similar tastes to the target user and recommends items that these similar users enjoyed. In the second case, the system recommends items similar to those that the target user has already liked, based on the preferences of other users. Graphs can be used to model relationships between users and items, as well as to identify communities of users with common interests.

The second approach, content-based models, focuses on the characteristics of the items themselves to generate recommendations. For example, a movie recommendation system might use information about genres, directors, actors, or themes to suggest movies similar to ones the user has already enjoyed. Graph algebra makes it possible to model the relationships between these characteristics and to determine the similarity between elements based on their attributes. Bipartite graphs, for example, can be used to connect items to their attributes, and graph similarity measures can be applied to identify the most relevant items to recommend.

It is also possible to combine collaborative filtering and content-based model approaches to create hybrid recommender systems. These systems take advantage of the strengths of each approach to provide recommendations that are more accurate and tailored to individual user preferences. The use of graph algebra in these hybrid systems effectively integrates information about users, items, and their attributes, providing a unified framework for analysis and generation of recommendations.

In this section, we will explore the different methods and techniques used in building recommender systems based on graph algebra. We will illustrate each approach with concrete examples and detail the advantages and disadvantages associated with each method. Additionally, we will examine how recent advances in the field of graph algebra have led to significant improvements in the quality and personalization of recommendations.

Collaborative filtering is an approach to recommendation based on the preferences and past behaviors of users. The basic idea is that if two users have shown similar preferences in the past, they are likely to have common interests in the future. Thus, by analyzing user preferences and finding people who share the same tastes, it is possible to generate recommendations for each of them. This can be classified into two main categories: memory-based methods and pattern-based methods. Memory-based methods directly use user preference data to calculate similarities between them and generate recommendations. Model-based methods, on the other hand, build a predictive model using training data to estimate user preferences.

An example of a memory-based method is user-user collaborative filtering. In this approach, the similarity between users is calculated using a measure of distance, such as Pearson's correlation (*) or Euclidean distance. Then, to generate a recommendation for a user, we identify the users most similar to him, and we assign them weights according to their similarity. Recommendations are generated by calculating a weighted rating for each item, based on ratings given by similar users.

User-user collaborative filtering can be easily represented as a graph. Users are represented by nodes, and the edges between nodes are weighted by the similarity between users. Recommendations are generated by exploring the graph to find nearest neighbors and weighting the items they rated based on similarity.

Another example of a memory-based method is item-item collaborative filtering. In this approach, the similarity between items is calculated based on the preferences of the users who rated them. Recommendations for a user are generated by identifying items similar to those they have already rated and weighting their similarity. Element-element collaborative filtering can also be represented as a graph, with elements as nodes and edges weighted by their similarity.

(*)Pearson, K. (1895). Note on regression and inheritance in the case of two parents. Proceedings of the Royal Society of London, 58(347-352), 240-242.

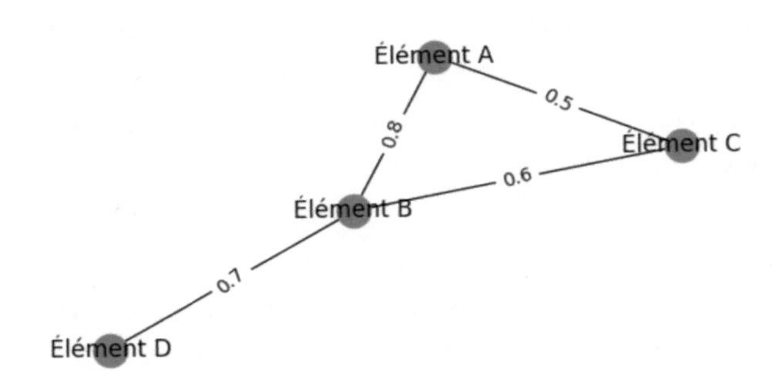

Here we have a graph illustrating item-item collaborative filtering. Each element is represented by a node of the graph. The edges between the nodes are weighted according to the similarity between the corresponding elements. The higher the similarity, the greater the weight of the edge. The graph visualizes similarity relationships between items, making it easier to identify items that are similar to those a user has already rated. These similar elements can then be used to generate personalized recommendations for the user.

As for model-based methods, a popular technique is matrix factorization, which decomposes the user-item preference matrix into two latent factor matrices. These factors represent hidden characteristics that explain user preferences. Recommendations are generated by reconstructing the preference matrix from the latent factors and identifying the items with the highest ratings for each user. Graph algebra can be used to represent latent factors as a bipartite graph and to compute preferences using algebraic operations on latent factor matrices.

Using graph algebra in collaborative filtering helps to better understand the relationships between users and items, improves the performance of recommendation algorithms, and makes it easier to visualize and interpret results. Additionally, algebraic methods can be easily extended and combined with other techniques to handle more complex problems and to include additional information, such as item metadata or contextual user information.

An interesting example of a practical application of collaborative filtering in the field of graph algebra is movie recommendation. Streaming platforms, such as Netflix or Amazon Prime Video, collect data about users' preferences based on the movies they watch, the ratings they give, and the wishlists they create. Using collaborative filtering methods based on graph algebra, these platforms can generate personalized recommendations for each user, taking into account the preferences of other users with similar tastes.

In summary, collaborative filtering is a powerful approach for recommender systems, which takes advantage of users' past preferences and behaviors to generate personalized recommendations. Methods based on graph algebra offer an interesting perspective and improve the performance of recommendation algorithms while facilitating their understanding and interpretation. In the context of our study of graph algebra and its applications, collaborative filtering provides a relevant example of how algebraic techniques can be applied to real-world problems and have a significant impact on our daily lives.

While collaborative filtering focuses on the relationships between users and items, content-based models leverage the intrinsic characteristics of items to generate recommendations. These models take into account information specific to each item, such as genres, authors, actors, keywords, etc., to determine the similarity between items and thus provide relevant recommendations to users.

Graph algebra can also be used to improve content-based models. One way to approach this problem is to construct a weighted graph representing the elements and their characteristics, where the nodes correspond to the elements and the characteristics, and the weighted edges indicate the strength of the relationship between an element and a particular characteristic. Then, algebraic techniques and graph theory algorithms can be applied to analyze and manipulate this graph, revealing hidden relationships and improving the quality of recommendations.

A common approach is to use the adjacency matrix of the graph to measure the similarity between elements. This similarity measure can be based on different methods, such as dot product, Pearson correlation or cosine distance. By multiplying the adjacency matrix by itself, one can identify indirect paths between elements and features, which helps uncover less obvious relationships and improves the quality of recommendations.

Another interesting aspect of content-based models is the ability to incorporate contextual information, such as user preferences based on time, day, weather, etc. To do this, we can build a multidimensional graph, where each dimension represents a different aspect of the context. Graph theory algorithms can then be adapted to account for these additional dimensions and generate contextual recommendations.

Take the example of recommending news articles online. Readers often have specific preferences for certain types of articles, such as sports articles, movie reviews, or political reporting. Using a content-based model, it is possible to determine the similarity between articles based on their characteristics, such as topics, authors, sources, and keywords. By constructing a weighted graph representing articles and their characteristics, and then applying algebraic techniques to analyze this graph, it is possible to generate article recommendations that match readers' interests.

In conclusion, content-based models are a complementary approach to collaborative filtering for recommender systems. By taking advantage of the intrinsic characteristics of elements and using graph algebra to analyze and manipulate this information, it is possible to improve the quality of recommendations and provide more relevant suggestions to users. Additionally, by integrating contextual information and leveraging advanced techniques from graph theory, content-based models can be adapted to generate contextual recommendations, providing an even more personalized user experience.

In addition to online news articles, content-based models can be applied to a wide variety of areas, such as recommending movies, music, products, or services. For example, to recommend movies to a user, one can take into account genres, directors, actors, ratings and other relevant characteristics. Similarly, to recommend music, one can use characteristics such as genre, artist, album, lyrics and other information related to the songs.

Graph algebra offers a powerful and flexible framework for enhancing content-based models. By exploring new techniques and algorithms derived from graph theory, researchers and practitioners can continue to refine and extend these models, in order to meet the challenges posed by modern recommender systems and provide ever more personalized user experiences. and relevant.

Overall, the application of graph algebra to content-based models illustrates how the concepts and techniques developed in this area can be used to solve real and complex problems. By continuing to explore the synergies between algebra and graphs, researchers and practitioners can discover new approaches and insights to address the challenges of data science and relational data processing, paving the way for important advances in many fields of application.

3) Bioinformatics and genomics

In this last part of the practical applications of graph algebra, we will approach the fascinating field of bioinformatics and genomics. These scientific disciplines are at the crossroads of biology, computer science and mathematics, and study biological processes using techniques of modelling, analysis and interpretation of data. Graphs play a vital role in exploring and understanding biological networks and molecular interactions, enabling researchers to decipher the mysteries of life and provide innovative solutions to medical and environmental challenges.

Bioinformatics and genomics have grown exponentially in recent years, notably thanks to technological advances in DNA sequencing and the increasing availability of large-scale biological data. These advances have broadened our understanding of the molecular and cellular mechanisms that govern living organisms, as well as revealing the genetic causes of many diseases and conditions.

However, the analysis of these large datasets requires sophisticated methods and tools to identify the complex relationships between biological elements and the underlying processes. Graph algebra offers a powerful theoretical and practical framework to address these challenges, allowing to model, quantify and analyze biological networks and molecular interactions.

In this part, we will look at two key applications of graph algebra in bioinformatics and genomics with sequence alignment.

Sequence alignment is a fundamental technique used to compare biological sequences, such as DNA, RNA, or proteins, to identify similarities and differences between them. This comparison can reveal valuable information about the evolutionary relationships between organisms, the structure and function of molecules, and the genetic causes of disease. We will explore graph-based sequence alignment methods, which take advantage of graph theory and algebra to effectively solve this complex problem.

Metabolic network analysis is an approach that aims to study chemical reactions and metabolic networks within cells and organisms. Metabolic networks can be represented as graphs, where nodes correspond to metabolites and edges represent chemical reactions. Graph algebra allows the extraction of key information about the structure, dynamics and regulation of these networks, as well as the identification of potential targets for therapeutic interventions or genetic modifications.

By exploring these applications, we will show how graph algebra can offer new perspectives and innovative solutions in the field of bioinformatics and genomics. The examples and case studies presented in this part will highlight the importance of interdisciplinary approaches and the need to combine knowledge in biology, computer science and mathematics to meet the challenges posed by large-scale biological data. Additionally, we will also discuss the challenges and opportunities of integrating graph algebra into bioinformatics and genomics research. Some of the main challenges include managing uncertainties and errors in biological data, validating graph models and analysis results, and adapting methods and algorithms to the specificities of biological systems.

Finally, we will conclude this part by discussing future prospects for the application of graph algebra in bioinformatics and genomics. We will examine emerging trends, such as the integration of machine learning and artificial intelligence in the analysis of biological networks, the exploitation of data for a more complete understanding of biological processes, and the development of new approaches. for viewing and interpreting complex graphs. These coming advances will help us make great discoveries in science and technology. They could change the way we treat illnesses, making them more suitable for each person. They could also help us modify genes to treat certain diseases, and understand how organisms use energy. This could have a big impact on many aspects of health and biology research. Sequence alignment is a crucial task in bioinformatics that involves comparing two or more biological sequences, such as DNA, RNA, or protein sequences, to identify similar regions and infer functional relationships. , structural or evolutionary. In this section, we will describe how graph algebra can be applied to sequence alignment to provide new insights and overcome some of the limitations of traditional approaches.

Sequence alignment methods are generally classified into two categories: global alignments, which attempt to compare the entirety of two sequences, and local alignments, which focus on regions of maximum similarity. Classical algorithms such as Needleman-Wunsch[1] for global alignments and Smith-Waterman[2] for local alignments are based on dynamic programming, an approach which, although powerful, can be time-consuming and memory-intensive to very long sequences or many sequences.

[1] Needleman, S. B., & Wunsch, C. D. (1970). A general method applicable to the search for similarities in the amino acid sequence of two proteins. Journal of Molecular Biology, 48(3), 443-453.
[2] Smith, T. F., & Waterman, M. S. (1981). Identification of common molecular subsequences. Journal of Molecular Biology, 147(1), 195-197.

Graph algebra offers interesting alternatives to traditional sequence alignment methods by representing biological sequences and their relationships in the form of graphs. One of the key concepts in this approach is the alignment graph, a bipartite graph where the nodes represent the positions of the sequences and the edges represent the potential correspondences between these positions. The objective is then to find a path in this graph which maximizes the sum of the correspondence scores of the edges while respecting the biological constraints, such as the conservation of the sequences or the absence of crossings in the alignments.

In the context of biology, DNA is a molecule that contains the genetic instructions used in the development, functioning and reproduction of all living organisms and some viruses. These instructions are encoded in the very structure of DNA, which is a long chain of four types of molecules called nucleotide bases: Adenine (A), Thymine (T), Cytosine (C), and Guanine (G). These bases form specific pairs (A with T, and C with G) to create the characteristic double helix of DNA. Let's take a simple example to illustrate the graph-based approach. Consider two DNA sequences, S1 = "ATCGA" and S2 = "ACGTA". Each letter in these sequences represents a nucleotide base: A for Adenine, T for Thymine, C for Cytosine and G for Guanine. We can construct a bipartite alignment graph with nodes representing the positions of the two sequences and edges connecting the nodes corresponding to identical base pairs. By searching for a path in this graph that maximizes the alignment score, we can identify the optimal alignment between the two sequences.

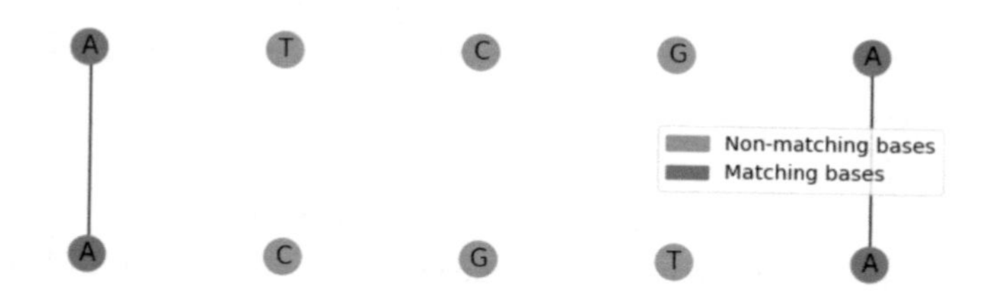

Graph algebra algorithms can be adapted to solve sequence alignment problems by exploiting the structure of the alignment graph. For example, we can use path-finding algorithms such as Dijkstra or Bellman-Ford to find the optimal path in the alignment graph, taking into account substitution scores and gap penalties. Additionally, graph partitioning methods can be used to divide the alignment problem into smaller subproblems, thus speeding up the alignment process for very long sequences or many sets of sequences. Here are the Dijkstra and Bellman-Ford algorithms for our example DNA sequence above.

```
Function BellmanFord(Graph, source):
    distance[] ← INFINITY
    distance[source] ← 0
    previous[] ← UNDEFINED

    for i from 1 to size(Graph)-1:
        for each edge (u, v) with weight w in Graph:
            if distance[u] + w < distance[v]:
                distance[v] ← distance[u] + w
                previous[v] ← u

    for each edge (u, v) with weight w in Graph:
        if distance[u] + w < distance[v]:
            return "Graph contains a negative weight cycle"

    return distance[], previous[]
```

```
Function Dijkstra(Graph, source):

    create vertex set Q

    for each vertex v in Graph:
        distance[v] ← INFINITY
        previous[v] ← UNDEFINED
        add v to Q
    distance[source] ← 0

    while Q is not empty:
        u ← vertex in Q with smallest distance
        remove u from Q

        for each neighbor v of u:           // only remaining vertices in Q
            alt ← distance[u] + weight(u, v)
            if alt < distance[v]:
                distance[v] ← alt
                previous[v] ← u

    return distance[], previous[]
```

Applying graph algebra to sequence alignment has several advantages over traditional approaches. First, it allows to deal with more complex alignment problems, such as the alignment of multiple sequences, by taking into account the evolutionary relations between the sequences and by exploiting the algebraic properties of the graphs to reduce the algorithmic complexity. Moreover, the graph-based approach is more flexible, as it can easily adapt to different biological constraints and to the specificities of the sequences studied.

However, applying graph algebra to sequence alignment also presents challenges. One of the main challenges is the management of uncertainties and errors in biological data, which can lead to errors in the construction of alignment graphs and erroneous alignment results. Moreover, validation of alignment results is a complex problem, because there is not always a "true" solution to compare sequences and graph-based methods can give different results than traditional methods.

Despite these challenges, graph algebra offers many future prospects for sequence alignment. The integration of machine learning and artificial intelligence into graph-based approaches offers opportunities to improve the accuracy and efficiency of existing methods. For example, graph-based deep learning techniques could be used to predict surrogate scores and gap penalties adaptively, based on local sequence characteristics and evolutionary relationships.

Chapter 6: Challenges and future perspectives

1) Scalability and processing of massive graphs

As data volumes increase, it becomes increasingly difficult to process and analyze large graphs efficiently. Massive graphs, such as social networks, biological networks or transport networks, can contain millions or even billions of nodes and edges, which poses major challenges in terms of storage, computation and visualization[1]. In this section, we discuss scalability issues and approaches to dealing with massive graphs using graph algebra and related techniques.

One of the main challenges in dealing with massive graphs is data storage management. Traditional data structures, such as adjacency matrices or adjacency lists, can quickly become memory inefficient for large graphs. Alternative approaches to graph representation, such as compressed graphs, sparse graphs, or hash-based graphs, can offer significant gains in terms of memory space and computational performance[2].

Regarding algorithmic aspects, many classical methods from graph algebra and graph theory may be inadequate to deal with massive graphs, due to their high temporal or spatial complexity. Parallel, distributed or approximate approaches may be needed to overcome these limitations and deal with large graphs efficiently. For example, depth or breadth-first traversal algorithms can be adapted to run in parallel on multiple processors or machines, while graph partitioning methods can be used to decompose massive graphs into smaller, more manageable subgraphs. .

[1] Newman, M. E. (2003). The structure and function of complex networks. SIAM review, 45(2), 167-256.

[2] Buluç, A., & Gilbert, J. R. (2011). The Combinatorial BLAS: Design, implementation, and applications. International Journal of High Performance Computing Applications, 25(4), 496-509.

Another major challenge in dealing with massive graphs is the visualization and interpretation of complex structures and patterns present in the data. Traditional graph visualization techniques, such as two-dimensional drawings or three-dimensional projections, may be inefficient or unreadable for large graphs[3]. Advanced visualization methods, such as dimension reduction techniques, adaptive layout algorithms, or hierarchical representations, may be required to effectively explore and analyze massive graphs. Moreover, the integration of machine learning and data mining methods can help to identify patterns or features in data, thus facilitating the interpretation and understanding of complex graphs[4].

In conclusion, dealing with massive graphs represents a major challenge and an exciting opportunity for graph algebra and related disciplines. Advances in storage methods, parallel or distributed algorithms, and visualization techniques can help overcome current limitations and enable efficient analysis and exploration of large graphs. Moreover, the interplay between graph algebra and other fields, such as machine learning, data mining, and cloud computing, can offer new insights and approaches to address scalability issues and performance associated with massive graphs.

[3] Herman, I., Melançon, G., & Marshall, M. S. (2000). Graph visualization and navigation in information visualization: A survey. IEEE Transactions on visualization and computer graphics, 6(1), 24-43.

[4] Chau, D. H., Kittur, A., Hong, J. I., & Faloutsos, C. (2011). Apolo: making sense of large network data by combining rich user interaction and machine learning. In Proceedings of the SIGCHI Conference on Human Factors in Computing Systems (pp. 167-176).

It is also important to emphasize the importance of interoperability between different methods and techniques for processing massive graphs. Researchers and practitioners should take care to develop open standards, data formats and exchange protocols to facilitate cooperation and sharing of results between different graph processing platforms and tools[5]. Additionally, designing intuitive and user-friendly user interfaces for graph processing software can help democratize access to these advanced techniques and encourage their adoption by a wider audience[6].

The rapid evolution of information and communication technologies, as well as the exponential increase in the volumes of data generated by sensors, social networks and computer systems, make the need to develop effective methods more imperative than ever. and scalable to process and analyze massive graphs. Advances in this area will have a significant impact on a wide range of applications, ranging from scientific research to online services, infrastructure management and political decision-making.

In sum, the scalability and processing of massive graphs is a growing field of research that requires the attention and expertise of the scientific and technical community. Advances in this field will open up new perspectives and opportunities to fully exploit the potential of graph algebra and related techniques in dealing with relational data, and help address the complex challenges and problems posed by graphs of large size and massive networks.

[5] Angles, R., & Gutierrez, C. (2008). Survey of graph database models. ACM Computing Surveys (CSUR), 40(1), 1-39.
[6] Von Landesberger, T., Kuijper, A., Schreck, T., Kohlhammer, J., Van Wijk, J. J., Fekete, J. D., & Fellner, D. W. (2011). Visual analysis of large graphs: Current state-of-the-art and future research challenges. Computer Graphics Forum, 30(6), 1719-1749.

2) Machine learning on graphs

In this section, we will discuss the growing importance of graph machine learning, an emerging field that combines graph algebra and machine learning methods to solve complex relational data processing problems[1]. Recent advances in this field have made it possible to develop innovative algorithms and techniques for the analysis, prediction and classification of data structured in the form of graphs[2]. Machine learning applied to graphs is different from traditional methods because it specifically uses the structure and the links of the graph. This helps uncover complex information and patterns that standard machine learning techniques, like linear regression or support vector machines, might struggle to capture.

Several challenges are associated with machine learning on graphs. First, data represented as graphs are often non-Euclidean (i.e., they do not follow traditional plane geometry) and high-dimensional, which makes it difficult to directly apply computational methods. traditional machine learning[3]. Second, machine learning algorithms must be able to deal efficiently with the structure, links, and attributes of graphs, as well as deal with noise, incompleteness, and complexity issues in relational data[4].

[1] Zhou, J., Cui, G., Zhang, Z., Yang, C., Liu, Z., Wang, L., ... & Sun, M. (2018). Graph neural networks: A review of methods and applications. arXiv preprint arXiv:1812.08434. [2] Wu, Z., Pan, S., Chen, F., Long, G., Zhang, C., & Yu, P. S. (2020). A comprehensive survey on graph neural networks. IEEE Transactions on Neural Networks and Learning Systems.
[3] Bronstein, M. M., Bruna, J., LeCun, Y., Szlam, A., & Vandergheynst, P. (2017). Geometric deep learning: going beyond Euclidean data. IEEE Signal Processing Magazine, 34(4), 18-42.
[4] Scarselli, F., Gori, M., Tsoi, A. C., Hagenbuchner, M., & Monfardini, G. (2009). The graph neural network model. IEEE Transactions on Neural Networks, 20(1), 61-80.

To address these challenges, researchers have developed various graph-specific machine learning approaches, such as matrix factorization methods, probabilistic graph models, and graph convolutional neural networks (GCNs)[1]. These techniques extract relevant features and representations from graphs, which can then be used to perform supervised or unsupervised learning tasks, such as node classification, edge prediction, and community detection. For example, matrix factorization methods, such as singular value decomposition (SVD) and non-negative matrix factorization (NMF), can be used to reduce the dimensionality of matrices associated with graphs (such as the matrix of adjacency or the Laplacian matrix) and to extract representative features of nodes and edges[2]. These features can then be used as inputs for classical machine learning models, such as support vector machines or random forests.

Probabilistic graph models, such as Markov random fields (MRFs) and Bayesian networks, provide a mathematical framework for representing and reasoning about uncertainties and dependencies present in graphs [3]. These models can be used to make inferences and predictions about node and edge attributes, as well as to learn the underlying structure of graphs from observed data[4]. Probabilistic graph models can also be combined with traditional machine learning methods, such as deep learning and ensemble methods, to create hybrid models that can effectively handle the complexity and variability of relational data.

[1] T. N. Kipf and M. Welling, "Semi-Supervised Classification with Graph Convolutional Networks," arXiv preprint arXiv:1609.02907, 2016.
[2] D. D. Lee and H. S. Seung, "Algorithms for Non-negative Matrix Factorization," Advances in neural information processing systems, vol. 13, pp. 556-562, 2001.
[3] J. Besag, "On the statistical analysis of dirty pictures (with discussion)," Journal of the Royal Statistical Society
[4] S. Li et al., "Graph Embedding Techniques, Applications, and Performance: A Survey," Knowledge-Based Systems, 2017

Graph convolutional neural networks (GCN) are another important approach for machine learning on graphs. Inspired by convolutional neural networks used for image processing, GCNs adapt convolution and pooling operations to data structured in the form of graphs. GCNs are particularly effective in extracting local and global features from graphs, taking into account the topological structure and the attributes of nodes and edges. The representations obtained from the GCNs can then be used to perform various learning tasks, such as node classification, edge prediction, and community detection.

Suppose we have a social network in the form of a graph, where each node represents a user and each edge represents a friendship connection between two users. Our goal is to predict if a new friendship connection will form between two users. We can use Graph Convolutional Neural Networks (GCN) to solve this task.[1] Here's how it works:

1. Graph construction: We represent the social network as a graph, where each node corresponds to a user and each edge corresponds to a friendship connection. Additionally, each node can have attributes such as age, location, number of friends, etc.

2. Node feature initialization: Each node is associated with an initial feature vector that can be based on its attributes. For example, if age is an attribute, each node could be represented by a vector where the first component is the age of the node.

3. Feature propagation: We apply a convolution operation on the features of the nodes to take into account the structure of the graph. The convolution is performed using a shared weight matrix for all neighboring nodes. It captures the local characteristics of the graph by aggregating the characteristics of neighboring nodes. Mathematically, the convolution can be formulated as follows:

$$h_v^l = \sigma\left(\sum_{u\in N(v)} \frac{1}{c_{u,v}} W^l h_u^{l-1}\right)$$

where h_v^l represents the feature vector of node v at layer l, N(v) is the set of neighboring nodes of v, $c_{u,v}$ is a normalization constant to compensate for varying degrees of neighboring nodes, is the weighting matrix at layer l, and is a nonlinear activation function.

4. Global aggregation: By using multiple convolution layers, GCNs can capture both local and global characteristics of the graph. Information from distant nodes in the graph can be aggregated across multiple convolution layers to obtain global node representations.

5. New connection prediction: Once we have obtained the final node characteristics from the convolution layers, we can use these characteristics to predict if a new friendship connection is formed between two users[2]. For example, we can use a binary classifier that takes as input the characteristics of two users and predicts the probability of forming a new friendship connection between them.

Graph convolutional neural networks (GCN) are therefore used to extract local and global features of graphs, taking into account the topological structure and attributes of nodes and edges. These representations can then be used for various learning tasks on graphs, such as node classification, edge prediction, community detection, etc. GCNs have shown good performance in many areas, including product recommendation, social media analysis, bioinformatics, chemistry, and many more.

[1] Michael M. Bronstein, Joan Bruna, Yann LeCun, Arthur Szlam, Pierre Vandergheynst. "Geometric Deep Learning: Grids, Groups, Graphs, Geodesics, and Gauges." arXiv:1611.08097 (2017).
[2] Jie Zhou, Ganqu Cui, Zhengyan Zhang, Cheng Yang, Zhiyuan Liu, Maosong Sun. "Graph Neural Networks: A Review of Methods and Applications." arXiv:1812.08434 (2018).

Machine learning on graphs has found many applications in various fields, such as bioinformatics, social networks, finance, medicine and physics. For example, in the field of bioinformatics, machine learning methods on graphs have been used to predict interactions between proteins, analyze metabolic networks and study relationships between organisms. In social networks, machine learning techniques on graphs have been applied to detect communities, analyze the propagation of information and model the processes of opinion and diffusion.

Despite the significant advances made in the field of machine learning on graphs, many challenges and research opportunities remain to be explored. For example, it is crucial to develop more efficient graph machine learning algorithms capable of dealing with large graphs of increasing complexity. Moreover, it is important to study the theoretical properties of machine learning methods on graphs, such as convergence, robustness and generalization. Finally, it is essential to design machine learning models on graphs that are interpretable and explainable, in order to facilitate the understanding of the results obtained and to promote their adoption in practical applications.

In sum, machine learning on graphs represents a promising area of research, which offers many opportunities to improve our understanding and processing of relational data. As graph machine learning techniques and algorithms continue to develop and mature, they can be expected to play an increasingly important role in solving complex and interdisciplinary problems, helping to shape the future of relational data processing and artificial intelligence.

Another promising aspect of machine learning on graphs relates to integration with other areas and techniques of artificial intelligence, such as natural language processing, computer vision and reinforcement learning. For example, machine learning methods on graphs could be used to analyze and model texts, images and temporal sequences by representing them as graphs, thus exploiting the richness of the relationships and dependencies between the constituent elements. [1]. Moreover, machine learning on graphs could be combined with reinforcement learning to develop intelligent agents capable of navigating and interacting efficiently in complex and structured environments in the form of graphs.

In a broader context, machine learning on graphs also offers interesting perspectives to address the ethical, social and environmental issues associated with the exploitation and analysis of relational data. For example, machine learning techniques on graphs could be used to study the biases and discriminations present in social networks, recommender systems and automated decision systems, identifying and correcting the underlying mechanisms responsible. of these phenomena[2]. Similarly, machine learning on graphs could be applied to the modeling and management of natural resources, ecosystems and infrastructures, helping to solve pressing problems such as environmental degradation, climate change and energy shortage.

In conclusion, machine learning on graphs constitutes an exciting and interdisciplinary field of research, which combines advances in graph algebra, machine learning and artificial intelligence to address the challenges and opportunities posed by relational data processing. As graph machine learning techniques and algorithms continue to develop and improve, they can be expected to play an increasingly important role in solving complex and complex problems. interdisciplinary, helping to shape the future of relational data processing and artificial intelligence.

[1] Wu, Z., Pan, S., Chen, F., Long, G., Zhang, C., & Yu, P. S. (2020). A Comprehensive Survey on Graph Neural Networks. IEEE Transactions on Neural Networks and Learning Systems.
[2] Brundage, M., et al. (2020). Toward Trustworthy AI Development. arXiv preprint arXiv:2004.07213.

3) Visualization and interpretation of complex graphs

The visualization and interpretation of complex graphs is a major challenge and an essential future prospect in the field of graph algebra and relational data processing. As we have seen throughout this book, graphs are natural structures for representing and modeling the relationships and interactions between entities in various contexts, ranging from social networks and recommender systems to bioinformatics and science. genomics. However, the increasing complexity and size of the graphs encountered in practice make it increasingly difficult to visualize and interpret them, thus limiting our ability to explore, understand and exploit the knowledge and information they contain.

To meet this challenge, several research and development directions can be considered. First, it is crucial to design and implement new techniques and methods of graph visualization that can effectively and comprehensibly represent the structures, patterns, and properties of complex graphs. This may involve using dimensionality reduction, clustering, and filtering techniques to simplify and summarize graphs, as well as exploiting advanced graphing techniques, such as 3D visualization, virtual reality, and reality. augmented, to improve the user experience and facilitate the exploration and analysis of graphs.

Then, it is important to develop approaches and tools for the interpretation of complex graphs, which allow to discover and understand the underlying mechanisms, processes and phenomena responsible for their structure and behavior. This may include applying pattern analysis and community detection techniques to identify groups, roles and functions of entities in graphs, as well as using machine learning and data mining techniques. to extract rules, patterns and predictions from graphs. Moreover, the interpretation of complex graphs can also benefit from integration with other fields and disciplines, such as psychology, sociology, economics and biology, which offer perspectives and theoretical frameworks to study and explain the phenomena observed in the graphs.

In summary, it is essential to foster collaboration and the exchange of ideas between researchers, professionals and users in the field of visualization and interpretation of complex graphs. To do this, we can organize events such as conferences and workshops dedicated to this field, as well as set up online platforms to share and disseminate data, software, tutorials and case studies. In addition, it is important to promote dialogue between people working on specific applications, such as social networks, recommender systems, bioinformatics, and developers of tools and methods for visualizing and interpreting complex graphs. . By adopting this approach, we will ensure that advances in this area respond directly to the challenges and problems encountered in practice.

In sum, the visualization and interpretation of complex graphs represent significant challenges and fascinating opportunities for the evolution of graph algebra and relational data processing. By developing new techniques for a better representation and understanding of graphs, by drawing on knowledge from other disciplines and by promoting collaboration between researchers, professionals and users, we can improve our understanding and use of complex graphs. These advances could have a major impact in many application areas, ranging from social networks and recommender systems to bioinformatics and many more.

Chapter 7: Conclusion

1) Summary of key concepts

During our exploration of graph algebra and its practical applications, we have covered a wide range of key concepts and techniques that are essential to understanding and fully exploiting relational data. It is appropriate, in conclusion, to review and briefly summarize the main points covered in the different sections of this book.

In the fundamentals of algebra and graphs (chapters 2.1 and 2.2), we introduced the key concepts of algebra, such as groups, rings and fields, as well as linear algebra. We also presented graph theory, with notions such as graphs, nodes, edges, types of graphs and properties, as well as representations of graphs. Next, we explored graph algebra (Chapter 3), discussing matrices associated with graphs, such as the adjacency matrix, incidence matrix, and Laplacian matrix. We also looked at algebraic operations on graphs, such as graph sum and product, graph power, and characteristic polynomials and spectrum.

In the chapter on Relational Data Processing Algorithms and Techniques (Chapter 4), we discussed graph mining methods, such as depth-first and breadth-first traversal, as well as Dijkstra's and Bellman-Ford's algorithms. for the shortest paths. We have also studied graph partitioning with spectral methods and flux methods, as well as graph optimization techniques, such as coupling algorithms, the shortest path problem and the traveling salesman problem.

We then applied graph algebra to specific practical areas (Chapter 5), examining social network analysis, with concepts such as centrality measurement and community detection, as well as recommender systems, with collaborative filtering and content-based models. Finally, we explored bioinformatics and genomics, addressing sequence alignment.
In Challenges and Future Perspectives (Chapter 6), we discussed the scalability and processing of massive graphs, machine learning on graphs, and the visualization and interpretation of complex graphs. These areas offer exciting opportunities for new

research and innovation, as well as for improving existing techniques and tools for processing relational data.

Throughout this book, we have stressed the importance of examples to illustrate and deepen our understanding of the concepts and techniques presented. Real-world examples not only ground abstract ideas in real-life situations, but also provide greater clarity and insight into how these concepts and techniques can be applied effectively and relevantly. Building on these examples, we were able to show how graph algebra can be used to solve complex problems and provide valuable insights in fields as diverse as social networks, recommender systems, bioinformatics and genomics. .

In summary, this book has sought to offer a comprehensive and in-depth overview of graph algebra and its practical applications, with an emphasis on understanding key concepts, exploring available techniques and tools, and presentation of various examples to illustrate the usefulness and impact of these ideas in real contexts. We hope that this exploration has not only provided you with a solid foundation to further explore these areas, but also inspired new ideas and perspectives for the future of relational data processing.

As technologies and data collection methods continue to evolve and expand, relational data will play an increasingly important role in our understanding and analysis of the world around us. The concepts and techniques presented in this book, while already powerful and versatile, are only the beginning. It is essential to continue to develop and refine these tools, and to apply them to new areas and problems, in order to take full advantage of the enormous potential that relational data offers.

Finally, in the spirit of curiosity and exploration that has guided this book, we encourage you to pursue your own journey in the study and application of graph algebra and relational data. Whether digging deeper into the concepts presented here, developing new techniques and methods, or applying these ideas to new problems and challenges, there is a world of opportunity to discover and explore in this exciting and ever-evolving field. .

2) Implications and impact on relational data processing

As we explored graph algebra and its applications, we shed light on the growing importance and relevance of this discipline in today's context of relational data processing. Rapid advances in data collection, analysis, and modeling have led to a proliferation of relational data, highlighting the need for effective methods and tools to process, analyze, and understand this data. . In this section, we wish to reflect on the implications and impact of graph algebra on relational data processing, highlighting areas where these concepts and techniques have already had a significant effect, and outlining challenges. and the opportunities that arise in the future.

First, graph algebra has allowed us to better understand and analyze the complex relational structures and networks that are ubiquitous in our modern world. Whether social networks, recommendation systems, biological networks or complex systems of interactions between entities, graph algebra provides a solid mathematical framework and powerful tools to study the properties, dynamics and the underlying mechanisms of these networks. Furthermore, graph algebra has played a vital role in the development of new techniques for analyzing relational data, which go beyond traditional methods of statistics and machine learning.

These techniques, such as spectral analysis, partitioning methods, and optimization algorithms, have solved complex problems and provided valuable insights into the structure, function, and properties of relational data.

Graph algebra has also helped facilitate collaboration and the exchange of ideas between different disciplines and areas of expertise. By providing a common language and a set of techniques applicable to a wide variety of problems and contexts, graph algebra has made it possible to form bridges between the fields of mathematics, computer science, physics, biology and social sciences, to name a few.

Finally, graph algebra has a big impact on how we approach and design systems and technologies that process relational data. Concepts and techniques developed in this area have led to significant innovations and improvements in areas such as data visualization, machine learning on graphs, massive graph processing, and modeling of complex systems. Despite these impressive achievements, there are still many challenges and opportunities in the field of graph algebra and relational data processing. Continuous progress in the collection and generation of massive relational data requires ever more efficient and adaptive methods and algorithms to manage this growing complexity. Furthermore, the exploration of new machine learning techniques on graphs, in particular those based on neural networks and deep learning approaches, offers considerable potential to improve our understanding and our ability to model complex relational systems. .

Another challenge is to develop graph visualization and interpretation methods that intuitively and effectively capture structures and relationships within relational data. Creating meaningful visual representations for large and complex graphs is a major challenge, but a breakthrough in this area could have a significant impact on our ability to explore and interact with relational data.

Finally, it is essential to continue to promote interdisciplinary collaboration and the exchange of ideas between researchers, practitioners and fields of expertise concerned with graph algebra and relational data processing. This will not only advance our understanding of the concepts and techniques of graph algebra, but also facilitate the application of these ideas to new problems and contexts, thus paving the way for new discoveries and innovations.

In conclusion, graph algebra has already had a profound and lasting impact on relational data processing, and it is clear that its importance will continue to grow as we strive to understand, analyze and model networks. and the complex relational structures that underlie our world. By rising to the challenges and seizing the opportunities that arise, we have the opportunity to shape the future of this exciting discipline and further expand its influence and impact across a wide range of fields and applications.

Made in the USA
Columbia, SC
22 November 2024